WHAT YOUR COLLEAGUES ARE SAYING . . .

"Who better than Pam Harris to help you introduce K–2 students to mathematical reasoning—the language, the music, and the poetry of mathematics. A must-read book filled with teaching strategies and creative ideas."

Jo Boaler
Nomellini & Olivier Professor of Education, Stanford University
Stanford, CA

"The abilities to count and to add are foundational to mathematics. All that follows is built upon these cornerstones. Get it right and math becomes 'figure-out-able.' In this book, Harris gives us the tools to get it right. Through real classroom examples, Harris takes us through strategies that are easy to adopt and effective in getting and keeping students engaged in the work of understanding mathematics."

Peter Liljedahl
Professor, Simon Fraser University;
Author, *Building Thinking Classrooms*
Vancouver, British Columbia, Canada

"It is with great enthusiasm that I endorse this transformative book. At the heart of this work is a compelling discussion of reasoning. Through rich narratives and classroom vignettes, we see that math fact fluency is not only figure-out-able but enjoyable—sparking curiosity and confidence in every student. In short, this book is a masterclass in making math fact fluency meaningful and accessible for all."

Dr. Nicki Newton
Math Consultant, Newton Education Solutions
Bridgeport, CT

"This book is a gift to primary teachers. It offers clear ideas we can use right away to help students build real understanding and develop as mathematical thinkers. From counting to additive reasoning, and through the power of models and Problem Strings, this book supports teachers in making instruction more purposeful and responsive."

Graham Fletcher
Math Specialist
Atlanta, GA

"Finally, the book that K–2 educators have been waiting for is here! Harris wrote a book that explores the complexity of foundational numeracy skills and shares research-based approaches to develop mathematical reasoning with our youngest learners. This book will not only help teachers cultivate curiosity and confidence and build a community of mathers, but it will also help teachers become the mathers they were always meant to be."

Deborah Peart Crayton
CEO & Queen Mather, My Mathematical Mind
Author, *Readers Read. Writers Write. Mathers Math!*
Charlotte, NC

"This book empowers K–2 teachers to deeply explore and understand the mathematics of the early grades, equipping them to foster students' mathematical reasoning and conceptual understanding. Grounded in research and filled with practical classroom examples, it offers valuable guidance for transforming early elementary math instruction."

Marria Carrington
Director of Mathematics Leadership Programs, Mount Holyoke College
South Hadley, MA

"What a powerful resource for equipping mathematics educators with the knowledge and skills necessary to develop their students' mathematical reasoning and promote their confidence as mathematicians!"

Janet D. Nuzzie
District Intervention Specialist, K–12 Mathematics,
Pasadena Independent School District
Pasadena, TX

"If you've ever heard Pam Harris talk about the Levels of Sophistication in Mathematical Reasoning and thought, 'I get it but what does that look like with MY students??' This book delivers! It reveals how young children develop counting skills and thinking strategies for mathematical operations. You'll gain insights into student development, practical tasks to showcase their thinking, and modeling techniques that benefit every child in your classroom."

Christina Tondevold
The Recovering Traditionalist
Founder, Build Math Minds
Orofino, ID

"Pam Harris offers teachers another invaluable resource for transitioning from repetitive procedural instruction to helping students develop deeper, more conceptual understanding of mathematics. Through numerous examples of student thinking about mathematical concepts in the K–2 classroom, she provides extensive treatment of the ways children approach problem solving. This comprehensive approach will certainly prove helpful for teachers who are developing their understanding of how students learn to make sense of mathematics independently."

John Tapper
CEO, All Learners Network Inc.
Burlington, VT

"This is a must-read for all K–2 teachers. Harris helps teachers understand how they can set their young learners up for long-term success in the math classroom. This book provides teachers with a resource that combines content knowledge, high quality instructional practices and ready-made instructional materials all in one! Upper elementary, middle school and even high school math teachers would all benefit from reading this book and better understanding the development of mathematical reasoning, too."

Brandon Pelter
Mathematics Teacher, Bridgeport Public Schools
Norwalk, CT

"This book is a game-changer for educators. With clear examples and practical strategies, it empowers teachers to transform their instructional practices. Packed with 'aha' moments and insights into the teaching and learning of mathematics, it inspires confidence in all of us. This is a must-read for anyone looking to make math instruction more meaningful for students while supporting the teacher with the 'why' behind it all."

Jennifer Lempp
Author and Educational Consultant
Alexandria, VA

"Math reasoning kicks algorithms to the curb when students engage their brains (not just their pencils) to solve problem strings using a variety of strategies. Harris takes the guess work out of teaching computation strategies intentionally by providing problem strings, teaching tips, and sentence frames for beginning and experienced teachers."

Carrie S. Cutler
Clinical Associate Professor of Mathematics Education,
University of Houston
The Woodlands, TX

"This latest book from Pam Harris takes research related to the major milestones of mathematical development in the primary grades and transforms it into a language that is easy to understand. Using carefully chosen examples, real world experiences, and student voices, Harris has written a book that is illuminating and practical. Primary educators, whether new to the profession or seasoned experts, will find ideas here that resonate and challenge them to listen closely in order to further thinking."

David Woodward
Founder and President, Forefront Education
Lafayette, CO

Developing MATHEMATICAL REASONING

The Strategies, Models, and Lessons to Teach the Big Ideas in Grades K–2

PAMELA WEBER HARRIS

Contributing Writers: **CAMERON HARRIS, KIM MONTAGUE, KOURTNEY LAMBERT PETERS**

CORWIN

FOR INFORMATION:

Corwin

A SAGE Company

2455 Teller Road

Thousand Oaks, California 91320

(800) 233-9936

www.corwin.com

SAGE Publications Ltd.

1 Oliver's Yard

55 City Road

London EC1Y 1SP

United Kingdom

SAGE Publications India Pvt. Ltd.

Unit No 323-333, Third Floor, F-Block

International Trade Tower Nehru Place

New Delhi 110 019

India

SAGE Publications Asia-Pacific Pte. Ltd.

18 Cross Street #10-10/11/12

China Square Central

Singapore 048423

Vice President and Editorial
  Director: Monica Eckman

Senior Acquisitions Editor,
  STEM: Debbie Hardin

Senior Editorial
  Assistant: Nyle De Leon

Production Editor: Tori Mirsadjadi

Copy Editor: Michelle Ponce

Typesetter: C&M Digitals (P) Ltd.

Proofreaders: Sarah Duffy and Wendy
                Jo Dymond

Indexer: Integra

Cover Designer: Gail Buschman

Marketing
  Manager: Margaret O'Connor

Copyright © 2026 by Corwin Press, Inc.

All rights reserved. Except as permitted by U.S. copyright law, no part of this work may be reproduced or distributed in any form or by any means, or stored in a database or retrieval system, without permission in writing from the publisher.

When forms and sample documents appearing in this work are intended for reproduction, they will be marked as such. Reproduction of their use is authorized for educational use by educators, local school sites, and/or noncommercial or nonprofit entities that have purchased the book.

All third-party trademarks referenced or depicted herein are included solely for the purpose of illustration and are the property of their respective owners. Reference to these trademarks in no way indicates any relationship with, or endorsement by, the trademark owner.

No AI training. Without in any way limiting the author's and publisher's exclusive rights under copyright, any use of this publication to "train" generative artificial intelligence (AI) or for other AI uses is expressly prohibited. The publisher reserves all rights to license uses of this publication for generative AI training or other AI uses.

Printed and bound by CPI Group (UK) Ltd, Croydon, CR0 4YY

Library of Congress Cataloging-in-Publication Data

Names: Harris, Pamela Weber author

Title: Developing mathematical reasoning. The strategies, models, and lessons to teach the big ideas in grades K–2 / Pamela Weber Harris; contributing writers Cameron Harris, Kim Montague, and Kourtney Lambert Peters.

Description: Thousand Oaks, California: Corwin, a Sage Company, [2026] | Series: Corwin mathematics series; vol 1 | Includes bibliographical references and index.

Identifiers: LCCN 2025019013 | ISBN 9781071967546 Paperback | ISBN 9781071929025 epub | ISBN 9781071928974 epub | ISBN 9781071928837 pdf

Subjects: LCSH: Logic, Symbolic and mathematical—Study and teaching | Algorithms—Study and teaching

Classification: LCC QA8.7 .H372 2026
LC record available at https://lccn.loc.gov/2025019013

This book is printed on acid-free paper.

25 26 27 28 29 10 9 8 7 6 5 4 3 2 1

DISCLAIMER: This book may direct you to access third-party content via web links, QR codes, or other scannable technologies, which are provided for your reference by the author(s). Corwin makes no guarantee that such third-party content will be available for your use and encourages you to review the terms and conditions of such third-party content. Corwin takes no responsibility and assumes no liability for your use of any third-party content, nor does Corwin approve, sponsor, endorse, verify, or certify such third-party content.

Contents

Preface	xi
About This Book	xvi
Language Use in This Book	xvii
Acknowledgments	xix
About the Author	xxi

PART I: SETTING THE STAGE — 1

1 MATHEMATICS FOR TEACHING — 3

What's the Purpose of Learning Math?	10
The Development of Mathematical Reasoning	11
Major Strategies	16
Conclusion	18
Discussion Questions	18

PART II: DEVELOPING COUNTING AND COUNTING STRATEGIES — 19

2 ALL ABOUT COUNTING — 21

The Difference Between Counting and Counting Strategies	22
Foundations of Number	23
How to Develop Counting	36
The Number Sequence in the Teens	37
The Number Sequence After the Teens	40
Meaning of Decades	41
Student Interview	42
Conclusion	43
Discussion Questions	43

3 COUNTING STRATEGIES — 45

- About Counting Strategies — 46
- Early Counting Strategies — 46
- The Counting On, Counting Back Strategy — 49
- Problem Types — 50
- Developing Counting Strategies — 58
- Conclusion — 66
- Discussion Questions — 66

PART III: DEVELOPING ADDITIVE REASONING — 67

4 THE MAJOR STRATEGIES FOR ADDITION WITHIN 20 — 69

- Additive Reasoning — 78
- Additive Strategies — 79
- Developing Addition Within 20 — 81
- The Get to 10 Strategy — 83
- The Next Two Major Strategies — 87
- The Using Doubles to Add Strategy — 88
- The Add 10 and Adjust Strategy — 92
- Comparing the Single-Digit Addition Strategies — 97
- Conclusion — 100
- Discussion Questions — 100

5 THE MAJOR STRATEGIES FOR SUBTRACTION WITHIN 20 — 103

- Developing Subtraction Within 20 — 105
- The Remove to 10 Strategy — 108
- The Next Two Major Strategies — 113
- The Using Doubles to Subtract Strategy — 114
- The Remove 10 and Adjust Strategy — 119
- Finding the Distance/Difference Strategy — 124
- Comparing the Single-Digit Subtraction Strategies — 125
- Conclusion — 126
- Discussion Questions — 126

6 THE MAJOR STRATEGIES FOR DOUBLE-DIGIT ADDITION — 129

- Developing Multi-Digit Addition Strategies — 131
- The Splitting by Place Value Strategy — 133
- The Next Two Major Strategies — 139
- The Add a Friendly Number Strategy — 139
- The Get to a Friendly Number Strategy — 144
- The Add a Friendly Number Over Strategy — 149
- The Give and Take Strategy — 153
- Comparing the Major Addition Strategies — 158
- Conclusion — 162
- Discussion Questions — 162

7 THE MAJOR STRATEGIES FOR MULTI-DIGIT SUBTRACTION — 165

- Developing Multi-Digit Subtraction Strategies — 175
- The Remove by Place Value Strategy — 177
- The Next Two Major Strategies — 178
- The Remove a Friendly Number Strategy — 179
- The Remove to a Friendly Number Strategy — 183
- The Remove a Friendly Number Over Strategy — 187
- Finding the Distance/Difference Strategy — 190
- The Constant Difference Strategy — 192
- Comparing the Major Strategies for Multi-Digit Subtraction — 196
- Conclusion — 199
- Discussion Questions — 199

PART IV: PUTTING IT ALL TOGETHER — 201

8 TASKS TO DEVELOP MATHEMATICAL REASONING — 203

- Sequencing Tasks — 204
- Problem Strings — 233
- Other Instructional Routines — 241
- Games — 250
- Hint Cards — 251
- Conclusion — 252
- Discussion Questions — 252

9 MODELING AND MODELS — 253

- Strategies Versus Models — 256
- The Many Meanings of *Model* — 260
- Exploring Models by Their Best Uses — 267
- Our Modeling Framework — 276
- Conclusion — 277
- Discussion Questions — 277

10 MOVING FORWARD — 279

- Mentor Mathematicians — 280
- Where to Start — 281
- Conclusion — 287
- Discussion Questions — 287

References — 289

Index — 291

Note From the Publisher: The authors have provided video and web content throughout the book that is available to you through QR (quick response) codes. To read a QR code, you must have a smartphone or tablet with a camera. We recommend that you download a QR code reader app that is made specifically for your phone or tablet brand.

Videos may also be accessed at **mathisfigureoutable.com/dmrcompanionK-2**

Preface

Trying to learn real math in a fake math classroom is a lot like looking at an autostereogram.

Autostereograms are pictures made of colored dots that at first glance look like visual noise. If you have the skill to focus your eyes just right, that visual noise resolves into a three-dimensional image that will appear to float in front of the page.

Source: Adapted from https://en.wikipedia.org/wiki/File:Stereogram_Tut_Random_Dot_Shark.png with CC Attribution-Share Alike 3.0. Retrieved May 17, 2025.

In this example, focusing beyond the page about three inches will reveal the image of a shark. Some of you will see the shark immediately, but many of you will not. Some will struggle and eventually see the shark. But some of you will probably never see the shark.

Think of this autostereogram of a shark as learning math.

Some of you will say, "I have no idea what's going on. The teacher is calling this a shark, but it doesn't look like what I thought a shark is. They say to memorize these weird squiggles, and bits of dots, so that must be what a shark is—I guess I will memorize these five squiggles and this horizontal wave—that must be what a shark is. I'll memorize that so that when I see it later, I'll know it's a shark." This was me.

Others, without guidance from a teacher, have the natural inclination to very quickly focus their eyes to see the shark. "Right there is a shark; surely everyone can see the shark. It's obviously right there. I wonder why the teacher is talking about five squiggles and the horizontal wave. Weird. It's just a shark." This was my son.

Still others can't see the shark but aren't willing or able to play the game the first group does. The nonsensical memorization of lines and dots when there is supposed to be a *shark* isn't enough for them to hold on to. Math becomes increasingly stressful. Students become convinced they just "aren't math people."

Who can blame them? Not me, not when we tell them there is a shark in the water, and then when they say they can't see it, we just throw more and more confetti dots at them.

The following is part of the 3-D image seen when viewed correctly.

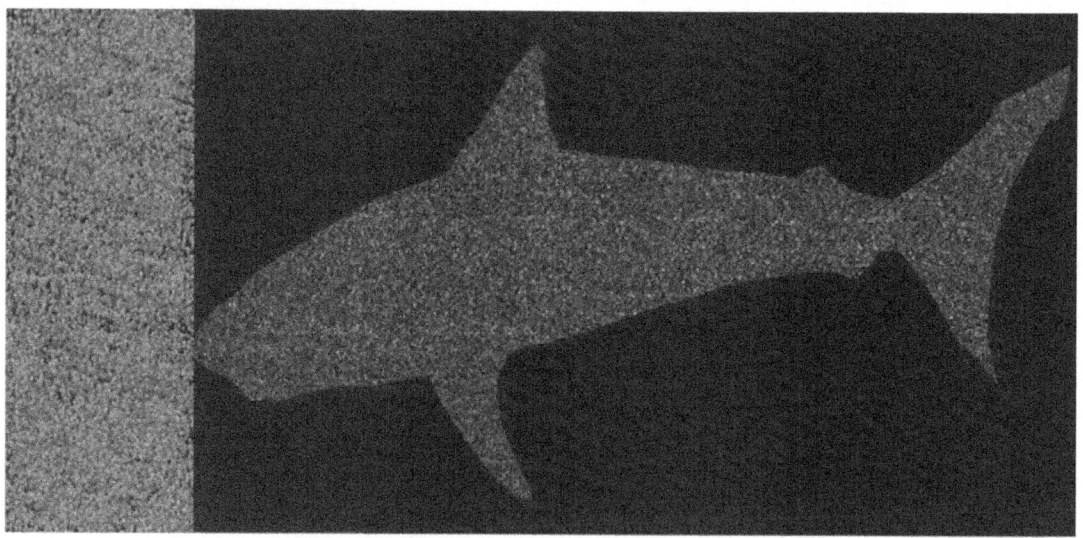

Source: Adapted from https://en.wikipedia.org/wiki/File:Stereogram_Tut_Random_Dot_Shark.png with CC Attribution-Share Alike 3.0.

A lot of really good work in the mathematics education field today focuses on getting students to be willing to try to focus on the shark again. *Building Thinking Classrooms* by Peter Liljedahl (2021) falls into this category. After years of nonsense, many students aren't willing to try to make sense of anything. Disruptions to the classroom norm like vertical-nonpermanent-surfaces and randomly chosen groups (as Liljedahl recommends) can do wonders. The willingness to try squinting (focusing) again is crucial.

But that's not enough. All students need guided, deliberate instruction aimed at how to see the shark. They need focused, purposeful instruction to do the mental actions that math-ers (Crayton, 2025) do. That's where this book comes in.

When I say math is figure-out-able, I'm saying everyone can be taught to see the shark. Everyone can be taught to see real math, to math their world.

When I say math is figure-out-able, I'm saying everyone can be taught to see the shark.

Some 25 years ago, as I was beginning to understand that math was figure-out-able myself, I sat down with my friend Mary. She was a second-grade teacher who taught near Austin, Texas. I had been reading about how mathematicians naturally have many ways to solve problems and an intuition for picking the approach that will be far easier and more efficient.

So I asked, "How do you think about adding 38 and 29?"

She added them by lining up the digits by place value in vertical columns, the traditional North American addition algorithm. Adding this way required "regrouping," handling digit overflow when the added digits come out to more than 9.

$$\begin{array}{r} \overset{1}{3}8 \\ +29 \\ \hline 67 \end{array}$$

I asked if she could think of any other way.

She looked at me as if I were an alien. Then, after a long pause, she lined up the numbers in columns again but with the order reversed.

$$\begin{array}{r} \overset{1}{3}8 \\ +29 \\ \hline 67 \end{array} \qquad \begin{array}{r} 29 \\ +38 \\ \hline \end{array}$$

Not quite what I meant. I took that to indicate that she had not yet had the opportunity to think about multi-digit addition outside the confines of the traditional algorithm. To her (and to young me) addition meant doing the steps. Excited to show her what I'd been learning, I began listing off other ways.

Judging by her rising panic, this was not helping.

The strategies I demonstrated were good strategies, but my teaching approach needed lots of work. Showing the strategies didn't give Mary any experience building the relationships. The strategies seemed like more algorithms Mary thought I was telling her to memorize. More lists of squiggles to see in an autostereogram. In other words, I discovered that giving a prescriptive list of strategies is a bad way to build the mental connections people need to actually use strategies as strategies.

In my early years of teaching, I started to realize that what I *thought* was math was only a shadow of what math really is.

The inkling that I was missing something massive had started when my eldest son—a first grader—started exploring math in ways I never had—even as a high school mathematics teacher. This drove me to dive into every bit of existing research I could get my hands on.

I began to realize that the math I had learned and the math I taught my high school students was more akin to trivia with extra steps than real mathematical reasoning. We memorized when to use which procedure and lists of steps, and we memorized the answers to pieces of those steps. Undoubtedly, some of my students were like my son and managed to piece together the logic and patterns behind those steps and really reason mathematically, but most of us, myself included, did not.

Some of you reading this are surprised that a high school mathematics teacher was not actually math-ing. Others are not surprised but are very frustrated. You know math makes sense, and has an underlying logic, but you don't understand why your teachers refused to teach you that logic. You might be wondering if the reason your teachers didn't teach you the underlying logic is because they didn't know that logic exists. Still others are like me, who bought into the myth that to do math was to rote-memorize and mimic. Students who tried to hold on to dozens of algorithms and formulas and, without really understanding why many of them work, try to pick the right one to get right answers.

If you are in this last group, you might also be starting to get an itch that math isn't quite what you thought it was. This can be very unsettling. It gave me something of an early mid-life crisis. Maybe you aren't in any of those groups. Maybe you know that math makes sense, that it is figure-out-able—that math taught as figure-out-able engages and enlightens. What you may not know is how to teach it that way.

If as a student you rocked at memorizing squiggles, then as a teacher you probably rock at teaching memorization of squiggles with rhymes, games, and/or drills. Your students have gotten the best possible start you could have given them. But most, like yourself, have not yet seen the actual shark in the water. You want to see the shark and help your students math. And now that you'll know better, you can do better.

If as a student you saw the shark, powerful and attention demanding, then as a teacher you probably started out confused about why most of your students didn't see what you saw. Maybe you found ways to help with that, maybe you didn't. Most likely driven by good intentions and frustration, you ended up teaching the same way you had been taught and accepting that only a rare few students would understand math the way you do. You want to help students see the beautiful sharks and are hungry for effective ways.

If as a student you had the looming, insistent feeling that there was something else in the water with you, and were occasionally freaked out that your teacher refused to acknowledge a fin cutting through the surf or explain the bite marks in your surfboard, then as a teacher you know something is up. If you knew the story about math your teachers were trying to sell you didn't add up, and were frustrated by their refusal to acknowledge that, then you want to improve. So as a teacher you do your best, sharing what you have been able to piece together. And you wish you could do more.

Whatever group you may fall into, this book will help you teach more real math to more students.

Even twenty-five years ago there was much research on this subject, but finding the good stuff and figuring out how to piece it together took time, active investigation, experimentation, and a background in higher math to sort the usable from the less useful.

I've found the single most powerful way to get students willing to squint (focus) again is to provide them with an experience where they see the shark. Where their squinting (focusing) is rewarded.

Real math-ing consists of reasoning using connections, understandings, and relationships. Fake math-ing is memorizing disconnected sets of facts and mimicking procedures, where each adds yet another ball of confusion to be juggled on top of the last one.

Real math-ing consists of reasoning using connections, understandings, and relationships.

This book is the guide to getting Kindergarten, first-, and second-grade students seeing sharks. It is the result of the research and experimentation I have done in the last twenty-five years to learn how to build students' brains to do more math, rather than merely burden those brains with more disconnected sequences of steps.

ABOUT THIS BOOK

Chapter 1 summarizes the contents of the first book in this series, *Developing Mathematical Reasoning: Avoiding the Trap of Algorithms* (Harris, 2025a). While it is intended for that book to be read before this one, this chapter serves as an abbreviated launch point or refresher. In short, algorithms have numerous shortcomings when used as teaching tools.

Chapters 2 and 3 lay out the critical foundation for all mathematics found in counting and Counting Strategies. Understanding counting has relevance at all grade levels, as we are either teaching it or watching for students who have not yet grown to also use Additive Reasoning.

Chapters 4 through 7 focus on what is required to develop more and more sophisticated additive reasoning, moving through the major strategies for addition within 20, subtraction within 20, multi-digit addition, and multi-digit subtraction. These major strategies are the "what to do instead" of teaching algorithms.

Chapters 8 and 9 move on to discuss the "how" to teach of Chapters 4 through 7's "what."

Chapter 8 details the different kinds of lessons to use.

Chapter 9 is about models and modeling, crucial tools and actions necessary to drawing out the patterns in student thinking needed for purposeful instruction in mathematical reasoning.

Chapter 10 concludes with how to begin in the classroom. Where to start and with what.

Each chapter includes tips and frequently asked questions throughout as well as actions the reader can take—either personal exercises or things to try in class.

Corwin and I will be publishing three additional grade-specific companion books (3–5, 6–8, and 9–12) on a sixth-month cadence once this book is released, which will offer more ideas, more practice, and more practical advice, concentrated specifically on each grade band. These books will be complementary to the anchor volume and this book.

Neither this book nor the rest of the (currently!) planned series have a focus on geometry, data, or measurement. Due to time and space constraints, I choose to restrict this series to the fundamentals of mathematical reasoning, as they are foundational to meaningful geometry, data, and measurement.

LANGUAGE USE IN THIS BOOK

There are a few terms that will be helpful to parse out before beginning.

- Mathematical Reasoning

    As used in this book, the term *mathematical reasoning* does not mean just a general ability to think. This is not a fuzzy "think better" approach that doesn't include doing the math and getting results. Mathematical reasoning is about building stronger brains and expects more, not less from students, giving them the tools to be successful at math-ing. It demands increasing sophistication of strategy. We meet students where they are and then develop from there. For example, students will not only know their addition facts, they will actually own them and be able to use the relationships in problems. It includes content-specific milestones such as understanding of place value, addition and subtraction, and so forth.

- Sophistication

    Sophistication as used in this book is a descriptor of how different levels of mathematical reasoning relate to each other. This includes how the domains of reasoning relate to each other, such as Additive Reasoning being more sophisticated than counting but also how individual strategies within a domain relate to each other. For example, both the Get to 10 and the Give and Take major strategies use Additive Reasoning. However, the latter requires more simultaneity and developed understanding.

    Sophistication is always a descriptor of a thought process, never a descriptor of a thinker.

- **Problem Strings**

  Problem Strings are everywhere in this book, because Problem Strings are the single best teaching routine for building sensemaking and teaching the major strategies. Problem Strings are deliberately ordered sequences of problems designed to develop an important big idea, model, or strategy. They are always meant to be teacher-led routines, with the teacher facilitating learning by revealing each problem one at a time, drawing out—modeling—student thinking, and crafting class discussions after each problem to highlight the patterns at work for the given strategy.

  For a more detailed look at Problem Strings for each grade level, and more than 200 example strings each, see *Numeracy Problem Strings: Kindergarten* (Harris 2025c), *First Grade* (Harris 2025b), and *Second Grade* (Harris 2024). Problem Strings will also be discussed in greater detail in Chapter 7.

- **Strategy and Model**

  The term *strategy* can mean instructional strategy or general problem solving strategy, but in this book, strategy means how you use mathematical relationships to reason through a problem. This is different than how you represent that strategy—that is a model.

  The term *model* is commonly used with many different definitions in mathematics teaching. In this book, model means "representing student thinking," which relates to the "representation of student thinking" and also as a "tool for thinking."

  For more on the differences between *strategy* and *model*, see Chapter 9. The full breakdown is reserved for the end of the book, as the context given by the proceeding chapters will aid greatly in understanding it.

Acknowledgments

- A mighty thanks to my son, Cameron Harris, for helping me find real math-ing and write this book.

- Thank you to Kim Montague and Kourtney Lambert Peters for also helping write this book—your expertise and attention to detail are greatly appreciated. It was fun! Thanks, Kim, for the great video footage! It's been fantastic working on three major projects with the two of you all at The. Same. Time! We mostly kept it all straight, right?

- Thank you to Sue (digits or numerals—that is the question!) and Kira and the rest of the gang at Math Is Figure-Out-Able for keeping the ship sailing while my head is down writing or I'm traveling. And thank you to Ann Latham for her meticulous work with permissions.

- Thanks to my husband, kids, grandkids, parents, and siblings who make life worth living and excellence worth striving for. And for the chocolate.

- A hearty thank you to Senior Acquisitions Editor Debbie Hardin and the Corwin crew for making this whole project work with a smile. Two down, three to go!

- Many thanks to Stephanie Lugo, Sarah Hempel, and Melisa Williams for inviting me into your classrooms to film. You and your students rock!

- And to the parents who allowed us to interview your three-, four-, and five-year-olds, I appreciate you and your kids!

With great respect and gratitude, I acknowledge the thousands of primary teachers and their students who have let me in their classrooms and taken our workshops—online and in person—and learned along with me. Your dedication to your craft is admirable. I hope you find this book helpful in your continued journey. You are my favorite group of teachers to work with because if I can convince you, you'll actually do it!

And to the author of life, thank you, Lord, for giving me a message worth sharing and the people around me to help me get it out.

PUBLISHER'S ACKNOWLEDGMENTS

Corwin gratefully acknowledges the contributions of the following reviewers:

Janet D. Nuzzie
District Intervention Specialist Pasadena Independent School District
Pasadena, TX

Jennifer Lempp
Author and Educational Consultant
Alexandria, VA

Brandon Pelter
Mathematics Teacher
Bridgeport Public Schools
Norwalk, CT

Carrie S. Cutler
Clinical Associate Professor of Mathematics Education University of Houston
The Woodlands, TX

Adina Rochkind
5th- and 6th-Grade Math Teacher
Baltimore MD

About the Author

Pamela Weber Harris is changing the way we view and teach mathematics. Pam is the author of several books, including the *Numeracy Problem Strings K–5* series, *Building Powerful Numeracy*, and the *Foundations for Strategies* series. As a mom, a former high school math teacher, a university lecturer, and an author, she believes everyone can do more math when it is based in reasoning rather than rote-memorizing or mimicking. Pam has created online *Building Powerful Mathematics* workshops and presents frequently at national and international conferences. Her particular interests include teaching real math, building powerful numeracy, sequencing Rich Tasks to construct mathematics, using technology appropriately, and facilitating smart assessment and vertical connectivity in curricula in schools PK–12. Pam helps leaders and teachers make the shift that supports students to learn real math because math is figure-out-able!

PART I
Setting the Stage

Chapter 1: Mathematics for Teaching

CHAPTER 1

Mathematics for Teaching

In the early 2000s, I facilitated a workshop for K–2 teachers near Austin, Texas. I couldn't help but notice that something had stirred the proverbial hornet's nest as the teachers came in. Everyone was mad enough that I knew we would struggle to get any meaningful learning done without clearing the air. So I asked them what was going on.

It turns out that a colleague of theirs, as part of the requirements of her university master's program, had administered an IQ test to each of them.

These teachers were upset because the math section of the test was "completely unfair." They thought the problems they were asked to solve, without being able to write anything down, were unreasonable.

> **TIP**
> Maybe it's not the best idea to use your colleagues as guinea pigs for your master's IQ test project. That may or may not go well.

They gave an example problem that involved "traveling 750 miles, going 65 miles per hour . . . and we don't even remember the rest because who can hold all of that in your head?!"

Something about the way they said "in your head" caught my attention, so I asked for exactly what was in their head when they heard this problem.

My question was not one they were expecting. To them, there was only one way to think about this problem so far. The digits 7, 5, 0 and then the digits 6, 5, and that it was unreasonable to expect them to hold five digits in their heads while trying to listen to the rest of the problem.

Because their sense of *number* was digits and their sense of *math*-ing was doing things with columns of digits, these teachers were not thinking about the relationship between about 750 and around 65. They weren't creating a mental map as the problem was being said, that 750 is a bit more than 10 times 65. It's not that they were refusing to consider the connections. It's as if those connections didn't exist.

This was a major epiphany for me. That, at no fault of their own, students (and teachers) could have such a limited conception of numbers.

Perhaps more importantly, working with those teachers the rest of that day I saw them develop more sophisticated mathematical reasoning. I witnessed again that anyone can *math*. Often people just don't know it's a thing. Those teachers didn't know there was an option besides lining up digits into columns and doing an algorithm.

I wished there was a book to help teachers understand the holes in their own mathematical reasoning and teach students so that those holes didn't get passed along.

It took a minute, but this is that book.

The mathematics of early grades is neither straightforward nor simplistic. As Dr. James Tanton (2019) observed, "Even the act of counting is fundamentally subtle and nuanced. There is so much intellectual richness to probe and explore there." This is true for the teaching of early grade mathematics. In Grades K–2, there are foundational concepts and benchmarks of reasoning that are not obvious to recognize or simple to teach, even if we ourselves have long since learned the content of those grades.

How can we do this teaching of real math-ing? Check out this first-grade classroom lesson.

Melisa, a first-grade teacher, starts the lesson by commenting that this Problem String is connected to their recent work where they used 10 and adjusted. Melisa asks Romie to repeat what the student had said previously.

Romie responds, "Tens are helping us with our nines."

Melisa says and writes the first problem, reminding students to signal when they have an answer with a thumbs up: 4 + 10.

"Addison, what is that one?" asks Melisa.

"Um, it equals 14."

Melisa responds as she draws a number line, "We've been working on that for a long time, haven't we?"

She asks, "Let's see how that can help us with the next problem: 4 + 9. Gabi, what do you think?"

Gabi responds with long pauses, "Um I think it equals . . . um . . ."

Melisa suggests that she look up at the first problem to see if the connection helps.

Still tentative, Gabi tries "Is it 13?"

Melisa gives a neutral response, not cueing Gabi if her answer is correct, "Ok, why do you think it's 13?"

Gabi, far more confident now, responds without pausing, "I think it's 13 because if you take away 1 from 14, it equals 13."

"Nicely done. That was really well explained," says Melisa, and she represents Gabi's thinking by drawing a new number line with a jump of 10 and then a jump back of 1. Melisa's words and modeling focus the students' attention on Gabi's thinking:

- Rather than "correct" or "right answer" Melisa says, "Nicely done," and "well explained."
- She makes Gabi's thinking visible on a model that provides Gabi with a visual-spatial representation that can assist her to reflect on her own thought process.
- This number line model makes Gabi's thinking (and the target strategy) more accessible to the class, and because her thinking is point-at-able, it's more discuss-able.

As Melisa draws the jump back 1, she asks, "And Gabi, why did we have to jump back one?"

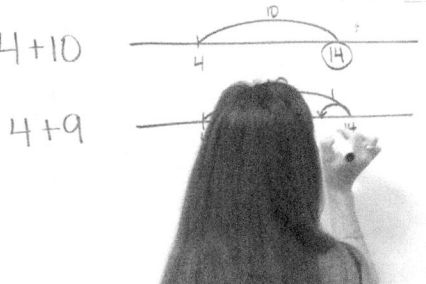

Gabi responds, "Because the 10 turned into a 9."

Melisa points to the number line: "Yes, it went down 1, didn't it? Because it was 10," pointing to the first equation, "and now it's 9," pointing to the second equation.

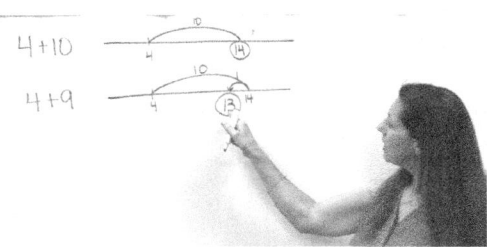

"You explained that one very well. Good job." Melisa moves on to the next purposefully planned question, "7 + 10. Thumbs up when you know this one."

You can almost see the gears turning, and then Jayson sits up tall with his thumb proudly up. Melisa calls on him, "Jayson, what is 7 + 10?"

Jayson answers, "17."

"Very nice," Melisa says as she draws a number line to represent the problem. To help students focus on the important relationships, she repeats the problem as she draws, "7 + 10, we've been working a lot on that one, haven't we? 7 + 10 is 17," and circles the 17.

She continues, "Let's see how 7 + 10 can help us with the next one. Some of you probably already know what the next one is going to be." She writes 7 + 9 on the board.

Melisa just gave two different nudges with two different goals:

- "How can it help with the next one" can help students potentially realize that they can use + 10 to help with + 9.

- "You might know what the next one will be" can help challenge those students who might have already been using the strategy. They get a chance to predict what the next problem could be based on the pattern, giving the rest of the students space to solve the problem.

Melisa asks, "Ben?"

Ben responds, "Uh ... 16."

"Why?" Melisa prods.

Ben smiles, "Because, like, 7 plus 10, but 1 less."

As Melisa represents his thinking, she says, "So 7 plus 10, that's what our first problem was." She points to the number line above, "and that was 17. But you said, we had to go ..."

Ben finishes, "Back one."

As she draws the jump back one, Melisa asks, "Why did we have to go back one again?"

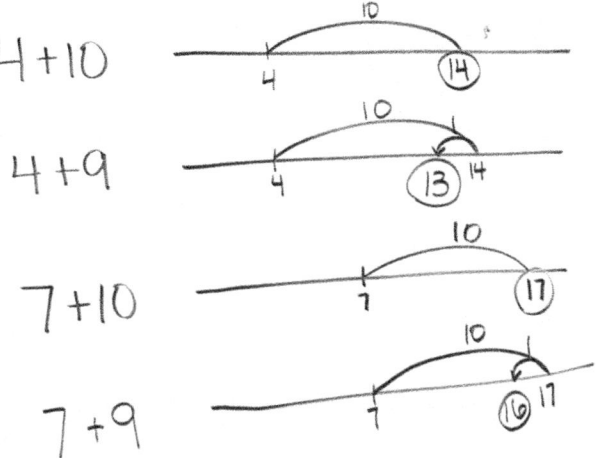

Ben answers, "To make 16."

"And also, this one was 9, right?" Melisa points to the current problem, "And this one was 10, and we had to go back 1 to get our 9, right?"

When Ben responds "yes," Melisa changes the pattern by asking a problem without a helper first. "Okay, good job, let's check out this last one, last one for this string." And she writes: 8 + 9.

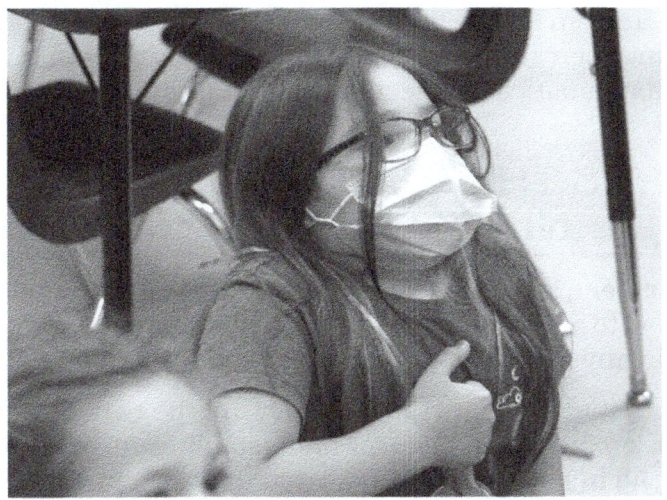

After seeing many thumbs up, Melisa asks, "Troy, what do you think?"

"Uh, 8 + 9 equals..." Troy begins, "If it was a 10, it'd be 18, but, it's a 9. It would just be taking 1 away. And if you take 1 away, it would be 17."

Melisa restates his words while she draws a number line to represent his thinking, "So you said, if it was 10, that would be 18, but we have to take away that 1 because it's actually a 9 we are adding.... So we landed on 17. Wow. look at that. You did so great with that!"

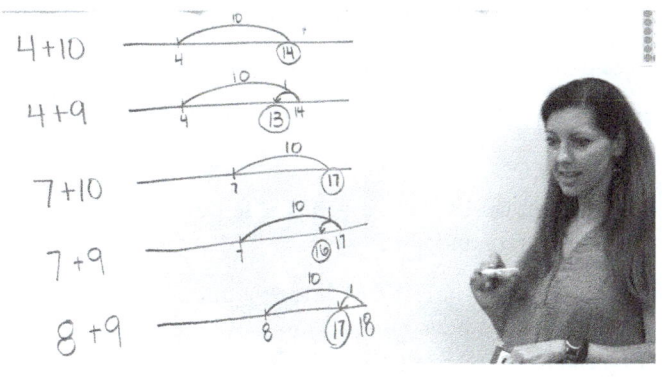

Watch this clip to see reasoning in Melisa's classroom.

https://qrs.ly/3zgl252

To read a QR code, you must have a smartphone or tablet with a camera. We recommend that you download a QR code reader app that is made specifically for your phone or tablet brand.

In this Problem String, we see students concentrating, smiling, justifying their thinking. As important, they are building and using mathematical connections. These students are *math*-ing.

WHAT'S THE PURPOSE OF LEARNING MATH?

For a moment, leave the world of education as it currently exists. If you had to argue for the inclusion of mathematics in K–12 education in any form, what arguments would you make?

I argue that mathematics and literacy are the two most powerful tools a person can have for understanding, organizing, and making an impact on the world. From there, a number of consequences logically flow. First, nothing, no matter how much it uses the trappings of mathematics, is worth spending valuable classroom time on if it doesn't aid in a student's ability to understand, organize, and make an impact on the world. Second, among possible ways to spend limited classroom time, priority must be given to those activities and approaches that make the greatest impact.

> *I argue that mathematics and literacy are the two most powerful tools a person can have for understanding, organizing, and making an impact on the world.*

Which means, third, that in the best-case scenario, algorithms and step-by-step procedures as answer-getting tools are near the very bottom of the barrel of ways to spend classroom time. In the worst case, teaching to mimic the steps of algorithms can actively inhibit many students' ability to understand, organize, and make an impact.

Let's return to the opening hypothetical to illustrate why. If your argument for the inclusion of mathematics in K–12 education is that students need to be able to get the answers to math problems as fast and easily as possible, math education ceases to need to be a pillar of the K–12 experience and can instead be reduced to a few weeks covering how to use a calculator.

FREQUENTLY ASKED QUESTIONS

Q: But Pam, that's what math is. The definition of math is to rote-memorize and mimic steps.

A: If you're thinking that, it's understandable. I invite you to consider that you might have been trapped by algorithms. Take heart! This book will help you get free.

THE DEVELOPMENT OF MATHEMATICAL REASONING

Our goal in math instruction should be to help students develop *mathematical reasoning*, which includes content. This is not some fuzzy *think better* game. Instead it is helping students learn to logically reason using mathematical concepts, strategies, models, and properties. Reasoning mathematically is solving problems using what you know and, in the process, building more and more real math.

Reasoning mathematically is solving problems using what you know and, in the process, building more and more real math.

There is a hierarchy of reasoning domains defined by increasing levels of sophistication and simultaneity. I call it the *Development of Mathematical Reasoning* (Harris, 2025). This represents the high-level hierarchy of milestones students must progress through to develop reasoning-based proficiency.

To reason through a problem, a student must grapple with multiple levels of complexity simultaneously. As students develop their brains, they create schema—ways of structuring their mental maps so that they can focus on the big picture or drill down to the specifics. Because they own this relationship map, their ability to grapple with multiple things will increase, leading to more efficiency and understanding.

As students develop their brains, they create schema— ways of structuring their mental map so that they can focus on the big picture or drill down to the specifics.

Sophistication includes the level of simultaneity employed as well as the complexity of the mathematical ideas at play.

Consider the following example of increasing simultaneity and sophistication when a student starts to Count On. Typically, for a problem like 6 + 3, the student has been counting the first set, 6, counting the second, 3, then counting them all, 9. We call this Counting Three Times. To Count On requires that the student can conceive 6-ness, to start at 6, not needing to count up to 6. This means that the student is simultaneously considering both the 6 objects in the set and that the word or numeral "6" represents the cardinality of the set (the last number in the count represents the amount). Then the student counts on from 6: 7, 8, 9. How does the student know when to stop? They must simultaneously keep track of the count and when to stop. This is cognitively difficult and takes time and experience to develop.

THE BEDROCK: COUNTING STRATEGIES

When students are beginning and solving beginner problems, they use Counting Strategies (see Figure 1.1).

FIGURE 1.1 • The First Level of Sophistication in Mathematical Reasoning

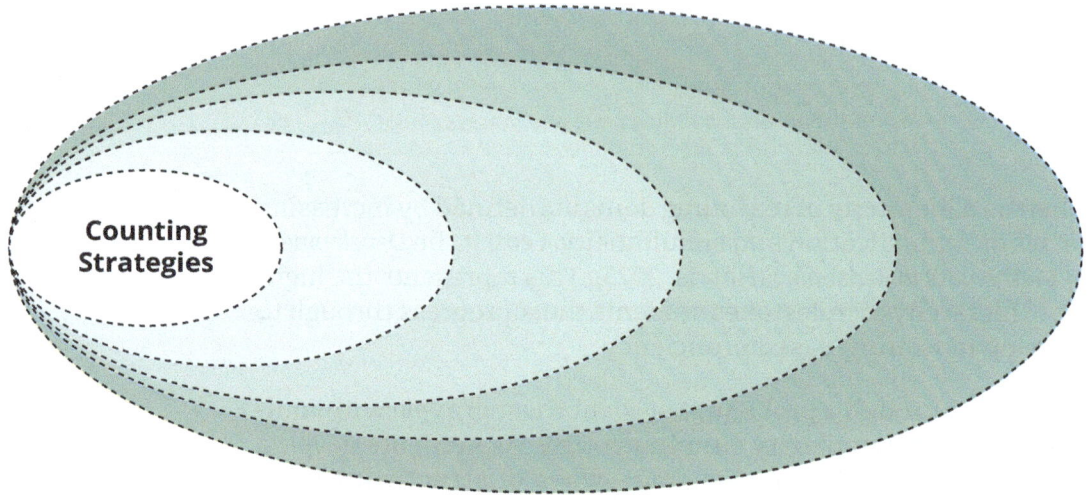

Source: Adapted from Math Is Figure-Out-Able at https://www.mathisfigureoutable.com/ with CC Attribution-NoDerivatives 4.0 International License.

Using Counting Strategies entails solving problems with one-by-one counting. We can solve addition, subtraction, and even multiplication and division problems counting one by one.

Using Counting Strategies is more than just being able to say the counting sequence. It's about solving problems and, while solving, considering the numbers involved as sets of 1s.

The Counting Strategies domain is the foundation of everything else. Teachers of counting are key to students developing all the domains because they begin it all. This is so important and exciting.

You will learn about major counting principles and Counting Strategies and how to develop them in Chapter 2. Develop these important Counting Strategies with an eye to then building Additive Reasoning.

> **TIP**
>
> Even if you are a kindergarten or first-grade teacher, it is important that you realize that the next goal is Additive Reasoning. You will almost certainly have students who are ready to develop further, so knowing those landmarks and how to build them is important.

THE FOUNDATION: ADDITIVE REASONING

Additive Reasoning is characterized by thinking in bigger jumps of numbers than 1 at a time (see Figure 1.2).

FIGURE 1.2 • The Second Level of Sophistication in Mathematical Reasoning

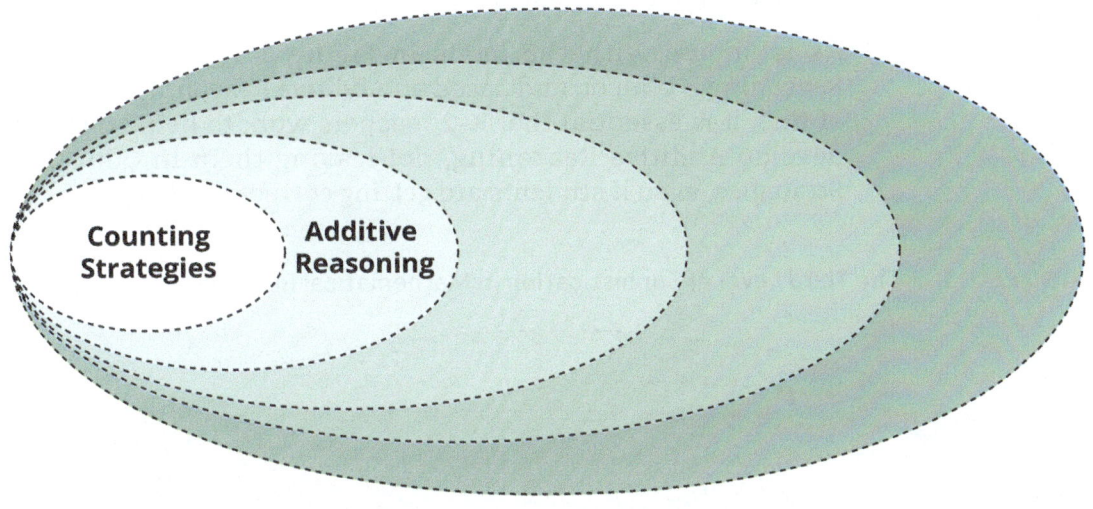

Source: Adapted from Math Is Figure-Out-Able at https://www.mathisfigureoutable.com/ with CC Attribution-NoDerivatives 4.0 International License.

An additive reasoner considers numbers simultaneously as sets of 1s *and* combinations of other sets of numbers. It's about composing and decomposing numbers in additive chunks. This Additive Reasoning is the major goal of kindergarten through second grade. As you help students develop counting and Counting Strategies, you are doing so with the end goal of those students developing Additive Reasoning. This means that you are always working toward students thinking in terms of bigger jumps than 1 at a time. This takes time, effort, and many experiences.

Grades 3, 4, 5, and 6 should continue to build Additive Reasoning with bigger, smaller (decimals), and more complicated numbers.

You will learn more about this important Additive Reasoning and how to develop it in Chapters 4 through 7.

THE NEXT DOMAINS: MULTIPLICATIVE, PROPORTIONAL, FUNCTIONAL REASONING

Multiplicative Reasoning is more mentally sophisticated than Additive Reasoning.

One is reasoning multiplicatively when considering the number of groups, the number in each group, and the total all at the same time. Solving a multiplication or division problem using Additive Reasoning looks like adding or subtracting one group at a time. Solving these problems with the more sophisticated Multiplicative Reasoning means using more than one group at a time, grouping the groups.

As is shown with the ovals in Figure 1.3, Multiplicative Reasoning is built on and based on Additive Reasoning. Because of this, it is essential that K–2 teachers work to help students develop Additive Reasoning, not leaving them in Counting Strategies, even if students are getting correct answers.

FIGURE 1.3 ● The Third Level of Sophistication in Mathematical Reasoning

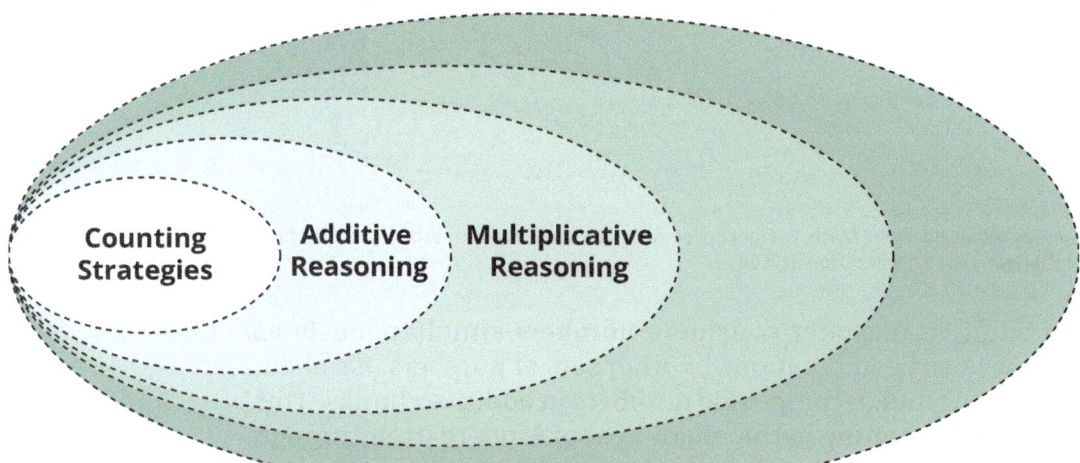

Source: Adapted from Math Is Figure-Out-Able at https://www.mathisfigureoutable.com/ with CC Attribution-NoDerivatives 4.0 International License.

In algorithm-driven instruction, this is the last stop for most students—there is just too much to memorize and keep straight if their understanding of math is mimicking procedures. It does not have to be. With the strong Additive Reasoning you will build in your K–2 students, they will have the sophistication of thought to be able to then develop Multiplicative Reasoning.

Proportional and Functional Reasoning are the domains of middle- and high-school mathematics.

As many as 90 percent of adults in the United States are not reasoning proportionally (Lamon, 2020). The vast majority of those same adults took classes in high school algebra, geometry, and more. This means they were getting answers, but because they did not have the necessary building blocks of reasoning, they were not able to reason more sophisticatedly.

If you are interested to learn more about these domains (Figure 1.4), read my *Developing Mathematical Reasoning: Avoiding the Trap of Algorithms* (2025).

FIGURE 1.4 ● The Full Spectrum of Mathematical Reasoning

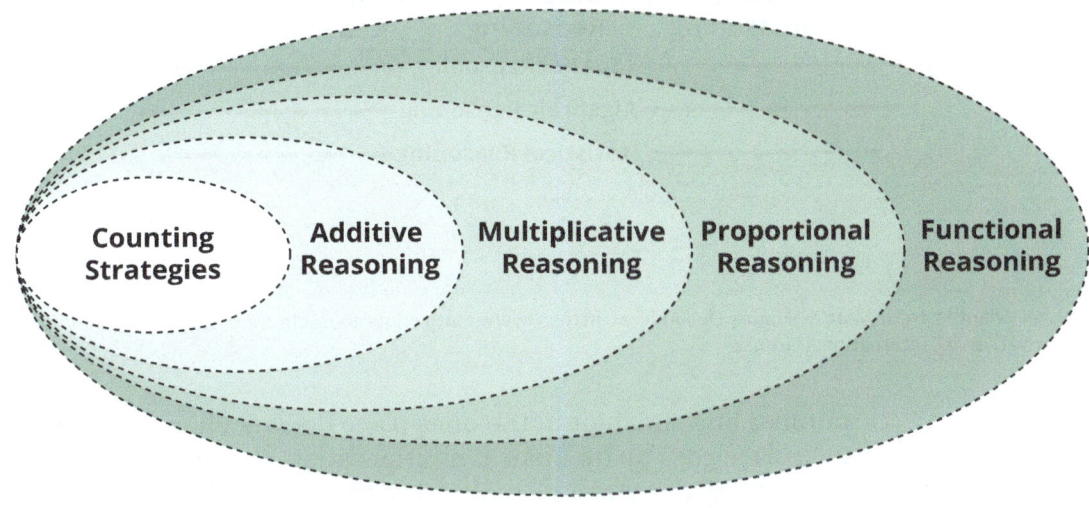

Source: Adapted from Math Is Figure-Out-Able at https://www.mathisfigureoutable.com/ with CC Attribution-NoDerivatives 4.0 International License.

SPATIAL, ALGEBRAIC, AND STATISTICAL REASONING

In addition to the five hierarchical domains in Figure 1.4, there are three longitudinal domains: spatial, algebraic, and statistical. I refer to them as longitudinal, because unlike the five hierarchical stages, they don't build off of other domains of reasoning in a specific sequence. Each of these three should be developed in concert with the hierarchical five (see Figure 1.5).

For example, spatial reasoning should see development all the way from Counting Strategies (1:1 tagging) to Multiplicative Reasoning (area models), to Functional Reasoning (distance and area defined by functions). Spatial reasoning is all about visual, geometric relationships of shapes, dimensions, measurement, location, graphs, and trends.

Algebraic reasoning is all about generalizing and working with generalizations. In counting, this might be that whether you have 5 dogs, 5 marbles, 5 balloons, or 5 anything else, the quantity is 5. In Additive Reasoning, this could be that you can add 10 to any

FIGURE 1.5 ● The Full Spectrum of Mathematical Reasoning With Longitudinal Domains

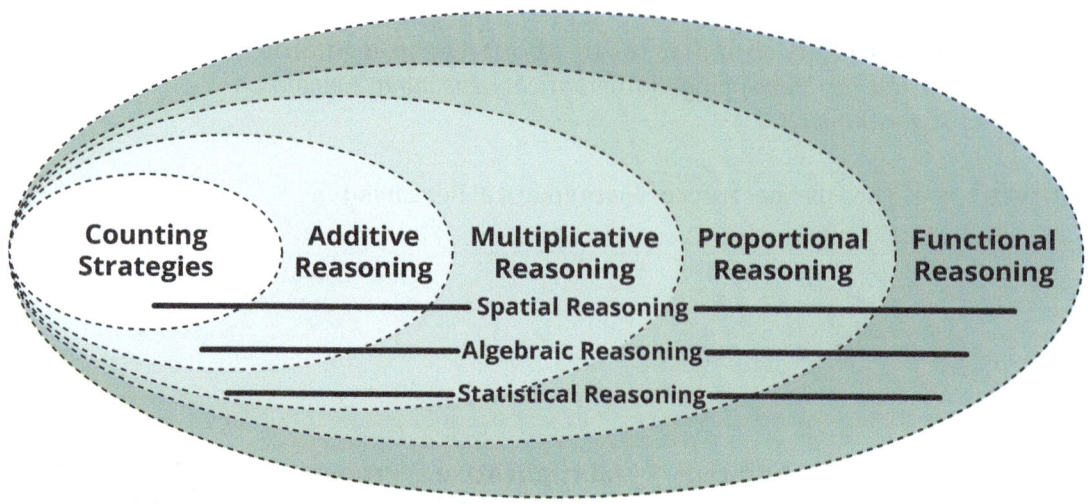

Source: Adapted from Math Is Figure-Out-Able at https://www.mathisfigureoutable.com/ with CC Attribution-NoDerivatives 4.0 International License.

number and the digit in the ones place doesn't change, and the tens place goes up by 1 ten. In multiplication, that might be that 5 multiplied by anything is half of that thing multiplied by 10.

Statistical reasoning concerns data, inferences, and predictions. This is the ability to draw useful information from data sets and create meaningful conclusions. It is also the ability to recognize when someone has manipulated factual data to favor their biased conclusions. At a basic level, statistical reasoning includes representing data in pictographs (the number of people in students' families), discussing the data (many of us have four people, some have less, and some have more), and possible changes (how would the graph change if we had students move in with five people in their family?).

Any lesson or routine intended to build reasoning must be conscious of how the material intersects the five hierarchical domains of reasoning with the longitudinal three. For example, spatial number line models for students learning Counting Strategies will not be very effective if those students do not yet have the spatial reasoning to appreciate the measurement meanings behind a number line.

MAJOR STRATEGIES

These are specific approaches to problem solving that take advantage of the human mind's natural pattern-finding abilities to build a student's mathematical understanding and empower them to solve problems quickly and efficiently. We can cultivate and train mathematical intuition that allows students to engage in *math*-ing.

Unlike algorithms, the learning of strategies is synergistic. Where each new algorithm is another series of steps to potentially misremember and confuse (is it carry forward or bigger bottom better borrow?), strategies are mutually reinforcing. The more relationships you own, the more strategies you learn, which in turn builds more relationships.

This book will help you learn the important counting principles and counting strategies, the major strategies for addition and subtraction within 20, the major strategies for double-digit addition and subtraction, and how to teach them.

Note that while, for example, there are four major strategies for addition where there is generally only one addition algorithm traditionally taught, it does not take four times as long to teach four strategies as it does to teach one algorithm. Each subsequent step of an algorithm breeds confusion or presents another opportunity for error, whereas every major strategy builds off of the one before it, bringing *increased* clarity and proficiency. Every subsequent major strategy learned makes mistakes with a previous one *less* likely.

> *Every subsequent major strategy learned makes mistakes with a previous one less likely.*

Finally, because strategies create relationships (through context-driven understanding), student retention of learning is far higher than that of algorithms—which tend to create disconnected, easily lost islands of procedure (Jensen & McConchie, 2020).

FREQUENTLY ASKED QUESTIONS

Q: Are you advocating for direct instruction or inquiry?

A: I'm advocating for a shift in goals—from mimicking algorithms to developing mathematical reasoning (which includes content). With that new goal in mind, how to teach becomes clearer: good guided inquiry for everything that is logical knowledge and clearly telling for the bit that is social/conventional knowledge. Teachers have clear mathematical goals; help students grapple *long enough*;

(Continued)

(Continued)

guide students to important generalizations through purposefully crafted discussions; anchor learning; and keep building on that learning to move the mathematics forward using tasks open enough that all students continue to have access and continue to be challenged. By doing this, students are solving problems not just correctly and efficiently but also more sophisticatedly. Students will be more successful longer. There will be more on how to do this in the rest of the book.

Conclusion

The purpose of math class is to develop mathematical reasoning, not mathematical answer-getting. What we need are not mere calculators but thinkers and do-ers of mathematics. Our role as teachers is not to have students rotely mimic algorithms that only provide answers to problems but to guide and support students as they develop as math-ers (Crayton, 2026). We can help students realize that they can use what they know to solve problems. Math is figure-out-able!

Discussion Questions

1. What was your experience as a student in math class? Did you more often reason through problems using what you know? Did you rote-memorize and mimic your teacher? How do you think that impacts the way you teach?

2. What's the difference between logical mathematical knowledge and social knowledge? Why does the difference matter?

3. What's the difference between an algorithm, a strategy, and a model? The book will further differentiate between these, but for now, how do these show up in your teaching?

PART II

Developing Counting and Counting Strategies

Chapter 2: All About Counting

Chapter 3: Counting Strategies

CHAPTER 2

All About Counting

FIGURE 2.1 ● The First Level of Sophistication in Mathematical Reasoning

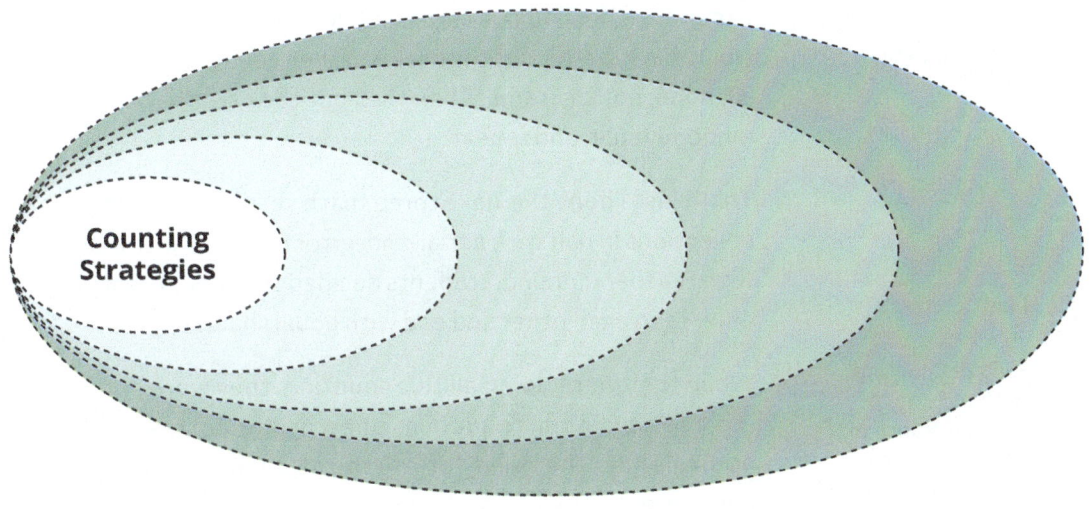

Source: Adapted from Math Is Figure-Out-Able at https://www.mathisfigureoutable.com/ with CC Attribution-NoDerivatives 4.0 International License.

It was the early days. I had been working with K–5 teachers in my children's classrooms. They were mostly keen to try things in my professional development sessions, but they kept asking, "What does this look like in classrooms with kids?"

To answer that, we decided to do a "lab" day, where I would do a lesson in each grade level.

I sat down with the kindergarteners with a card game I felt confident would be just right for the students at this time of year. I had the rules, supports, and differentiation planned and was ready to go.

> **TIP**
>
> Lessons will often not go exactly (or anywhere near) as planned, especially while you're learning how to teach real math-ing. That doesn't mean such lessons are failures! Understand why you are teaching a lesson, and then when you'll be able to pivot effectively when things don't go as planned.

> **TIP**
>
> The "Song of Counting" is a shorthand way to refer to the instance of students just learning to count not quite understanding what counting means. When a student "sings the song," they are repeating a sequence of words they don't understand or have context for. They don't yet understand that counting means determining a numerical value and that saying the names of the numbers in sequence is an organizational help to counting but not counting itself.
>
> Just like it is not enough for students to sing the song of the A, B, Cs and call it spelling, saying the counting sequence is not enough to constitute counting.

"Today we're going to play a fun math game," I began. "To start, you'll split the deck in half with your partner." I was all ready to continue with the game when the teachers looked at me with knowing grins.

Can you guess what happened?

We spent the rest of the time working together on splitting the deck in half!

It hadn't occurred to me that these five-year-olds would not have experience dealing out cards into two even piles. Instead, they would haphazardly cut the deck and then count each part: 25 and 27, nope, not the same! Then they put the cards back together, randomly split, and repeat.

That class spent the next three math sessions splitting other collections in half with a trial-and-error method, until gradually, with teacher nudging, students decided they could deal out the objects to each other and end with equal shares.

While they were doing all this counting, they were constructing the counting sequence, one-to-one correspondence, equivalence, the need for organization and keeping track, and more.

THE DIFFERENCE BETWEEN COUNTING AND COUNTING STRATEGIES

Counting is more than singing the song of counting. There's recognizing numerals but also having a sense of quantity. Knowing that one counting word correlates to each object, counting backward, and that number quantities are enveloped within each other—all of these and more combine to create a robust understanding of number and counting. These *things* have been called *counting principles* (Gelman & Gallistel, 1978). I'll refer to this set of knowledge as the *foundations of number*.

This chapter covers these essential foundations.

FOUNDATIONS OF NUMBER

The following principles are loosely listed in order of increasing sophistication, but these principles are all so intertwined and interrelated that you will rarely build any of them in isolation.

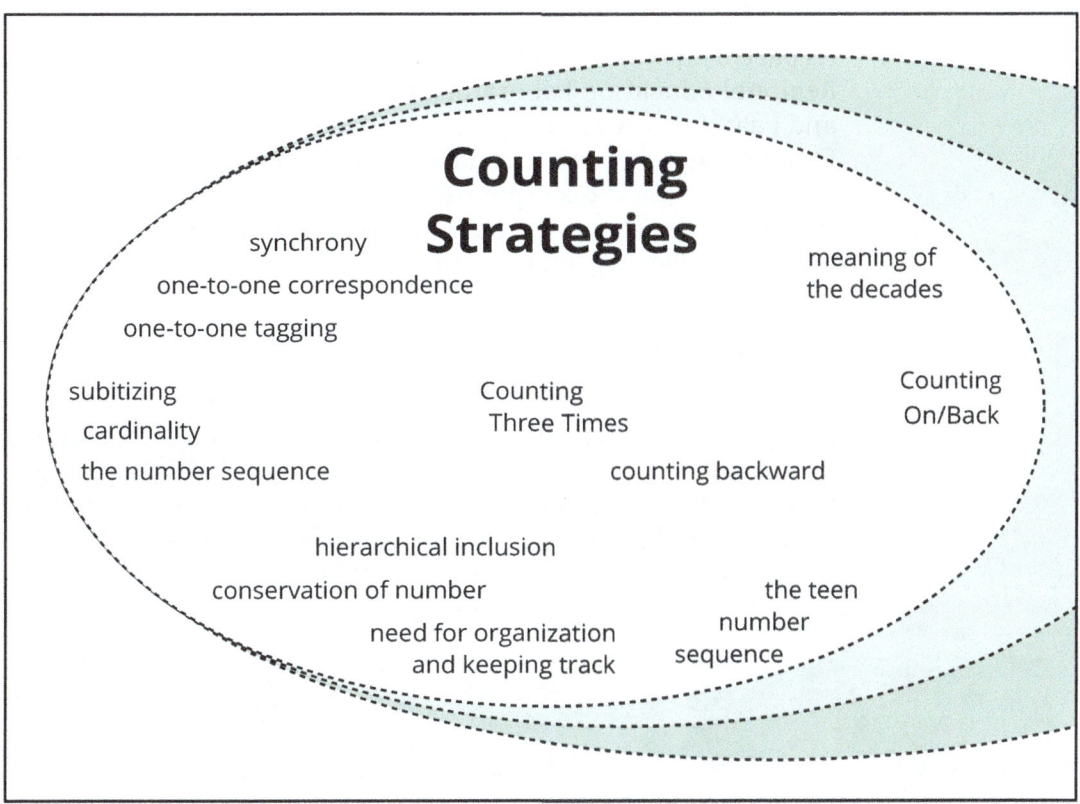

Catherine Twomey Fosnot, elementary mathematics researcher, explains,

> These landmarks in the mathematical environment epitomize children's struggles in their journey toward the horizon of number sense. These landmarks should inform our decision making as teachers because they characterize the shifts and steps in early mathematical development. Some of these landmarks may even become "horizons" in the journey toward numeracy. But it is important that they do not become goals devoid of context. They are not a series of flags or signposts to be reached and checked off, like a list of objectives. Instead, when we plan, we should try to design environments that are likely to enable children to mathematize in ways characteristic of the landmarks. The focus should be on articulating "the nature of the mathematical environment in which we hope students will eventually come to act" (Cobb 1997). (Fosnot & Dolk, 2001, p. 37)

A number of researchers in the math education space have identified more fine-grained nuances in counting principles and strategies than I describe here. I've deliberately limited what I include based on my own experience working with K–2 teachers to what will be the most directly and powerfully applicable.

As part of an initiative for teacher training with the Southeastern Regional Education Laboratories, my colleague Kim Montague and I sat down with individual students, interviewing them to find out what they can do and what they know of these counting principles. I'll use excerpts from these interviews to exemplify the principles, showing examples on the continuum of not owning yet, developing, and owning.

SUBITIZING

As I showed Rowan, a four-year-old, different numbers of fingers, he confidently called out the appropriate number, without counting for 2, 3, and 4 fingers, even with nontypical arrangements of fingers. Recognizing small amounts without counting is called *subitizing*, or the "direct perceptual apprehension of the numerosity of a group" (Clements, 1999).

Rowan called out "three" without counting.

Rowan called out "two" without counting.

Rowan called out "four" without counting.

Rowan called out "three" without counting.

Rowan called out "four" without counting.

When I showed Rowan six fingers (Figure 2.2), he paused, looked from left to right, paused, then said, "six." Because he was figuring out that there were 6 fingers, probably counting all six or counting on from 5; this is an example of not subitizing.

FIGURE 2.2 • Example of Not Subitizing

> **TRY IT**
>
> To help students subitize small numbers of objects, show an arrangement of 2, 3, or 4 objects, and ask them how many. If children need to count, allow them to count. Once they start to subitize and say the correct number, flash the objects by showing them briefly and then covering the objects or taking them out of view.

Subitizing can be thought of as a subset of unitizing, which is described on p. 34.

CARDINALITY

Knowing that the last number in a counting sequence represents the count is called *cardinality*. Children who have not yet constructed cardinality repeat the counting sequence when asked, "How many?" without realizing that the last number word they say represents "how many." When counting 5 objects, there's 5 objects, not 1, 2, 3, 4, 5.

When I showed Revan three fingers, he said, "three," demonstrating subitizing. When I asked him how he knew, could he count them, he tagged each finger as he said, "One, two, three." I said, "So, how many?" Revan repeated tagging and saying, "One, two, three."

> **TIP**
>
> Technically, children can subitize by recognizing the small amounts of 2, 3, or 4 with knowing the number word that represents that amount. For example, babies react differently to different numbers of objects (Clements, 1999). For our purposes, we'll refer to subitizing as perceiving the amount and saying the number word that represents that amount.

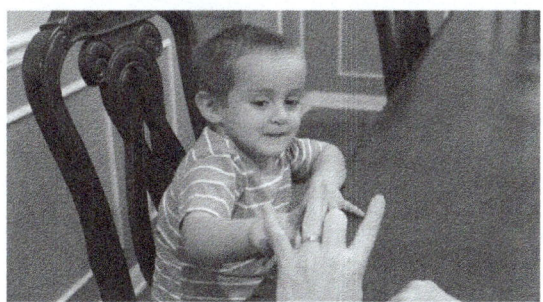

I asked, "So how many?"

A third time, he repeated tagging as he counted, "One, two, three."

I moved on, showed four fingers, and asked, "How many?"

He tagged each finger, "One, two, three, four."

I asked, "So how many?"

He paused briefly, looked at me and said, "Four."

I then showed him all ten fingers.

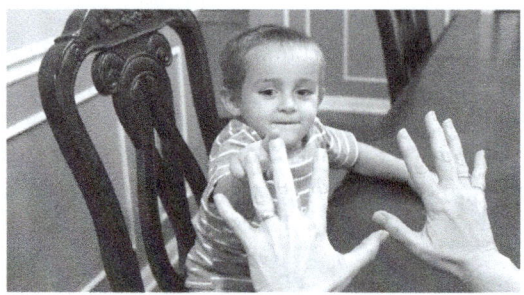

He immediately started tagging and saying, "One, two, three, four, five, six, seven, eight, nine, ten," and stopped.

I asked, "So how many?"

He paused, then said, "Ten."

Revan was constructing cardinality, realizing that when he is done counting ten objects, the answer to "How many?" is ten, that ten represents the total.

TRY IT

Help children develop cardinality! As they are counting to find an amount, listen to ascertain if they think they are done when they finish counting or if they correctly identify the last number in the count as the total. If they act like they are finished as they say the end of the count, make sure to follow that with, "So how many?" If the student continues to repeat the counting sequence but does not acknowledge the last number, say, "Ah, so there's five." As you interact with them during other counting opportunities, keep asking, "So how many?" When that student starts to reply with the last word, "five," you can keep building cardinality by reinforcing, "That's five" or "So, there are five."

FREQUENTLY ASKED QUESTIONS

Q: What do I do if my student just keeps saying the sequence of counting and does not end by repeating the last number?

A: Keep developing all of the other principles of counting. As you do, continue to repeat back that last number in the count, "So there's six." And make the count *count*, make it matter. Some students might need more experiences.

ONE-TO-ONE CORRESPONDENCE

One-to-one correspondence is a concept that encompasses other ideas like synchrony and strategies like one-to-one tagging. A student demonstrates that they own one-to-one correspondence when they count the amount correctly, can decide how many other objects are needed to pair with those in a set, or can infer that if each object in one group is paired with an object in another group the groups have the same number of objects (Chang & Fosnot, 2025).

> ### TRY IT
>
> Help students develop one-to-one correspondence by giving them experiences where it matters. Ask students to get enough pencils for their group or enough playground balls for the team. This means they need to determine how many students are in their group or the team and then get enough pencils or balls. Ask, "Do you have enough? Do you have too many? If there are 6 children, how many pencils do you need to get?"

SYNCHRONY

Using one counting word for each object when counting displays *synchrony* (Fosnot & Dolk, 2001). Using synchrony is a part of owning one-to-one correspondence. We see issues of synchrony when students can say the counting sequence, but they do not associate each number word with exactly one object. They might think that when someone asks, "How many?" their job is to sing the song of the counting sequence, but they don't realize that each counting word should be paired with exactly one object.

> ### TRY IT
>
> Help students build synchrony by playing Mr. Mixup (Clements & Sarama, n.d.). Mr. Mixup counts things but mixes up the count in some way. Introduce him to students, and ask students to listen carefully in case he mixes up his count. Sometimes have Mr. Mixup double count an object. Sometimes have him miss an object. Each time report the count and ask students if the count is correct. Have students explain the mixup and what is correct.

ONE-TO-ONE TAGGING

The strategy of touching each object with a finger or even with your gaze can be called *tagging*. If students accurately touch each object as they say one word, they demonstrate synchrony, and they are using one-to-one tagging.

Tagging is the action. If it's done one-to-one, it's called *one-to-one tagging*.

When Revan counted my fingers, he touched each of them one at a time in *synchrony*, demonstrating *one-to-one tagging*. When Rowan counted my fingers, he moved his eyes as he looked at each of my fingers one at a time and counted aloud one word for each finger. This also demonstrates one-to-one tagging. This is shown in the following images.

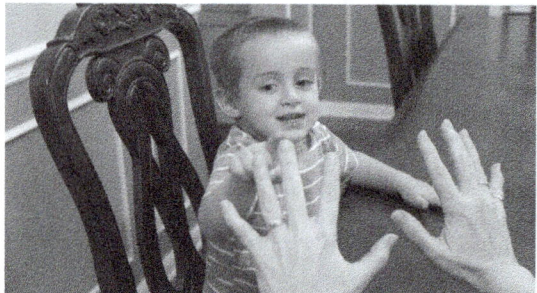 

If a student is doing the action of tagging by touching or pointing, but missing some objects or double-counting, they are tagging, but not one-to-one tagging.

TRY IT

You can encourage students to tag objects by demonstrating tagging when you count things. Use your finger to touch each object as you say each number word. Invite students to count with you as you tag each object.

HIERARCHICAL INCLUSION

Knowing and using *hierarchical inclusion* (Kamii, 1985; Piaget, 1965) means understanding that numbers nest within each other. If you have six objects, you also have five objects within that group.

Hierarchical inclusion is "the idea that numbers build by exactly one each time and that they nest within each other by this amount" (Fosnot & Dolk, 2001, p. 36).

When working with Rowan, he correctly counted seven blocks. Then he put them in the box.

After the blocks were in the box, I asked, "How many blocks are there in the box?"

Rowan answered, "Seven," showing conservation of number.

Then I asked, "If I take one out for you, so you can play with it, how many blocks are still in here?"

Rowan confidently answered, "Six."

I asked, "How do you know?"

He replied, "Because, I counted to seven, and that's how I know that there's six." Rowan understood that six is nested within seven.

He repeated this successfully as I took more blocks out, one at a time, reporting how many blocks were still in the box.

TRY IT

Help students develop hierarchical inclusion by asking questions about nested amounts. This can be experiences where students establish a total and wonder about 1 less than the total. For example, show a student 6 blocks. After the student has determined there are 6 blocks, remove one block, and ask how many blocks are left. If the student can correctly reason that 5 comes before 6 and is nested in 6 so that 1 less than 6 is 5, that student is showing understanding of *hierarchical inclusion*.

FREQUENTLY ASKED QUESTIONS

Q: Do I need to create lessons to teach each of these important counting principles, one at a time?

A: No, these strategies and big ideas are all interrelated. Students will best learn these principles not in isolation but as students solve problems. I am explaining them all individually so that you can recognize them, know how to test for them, and then you can tailor the way you interact with students based on that information.

CONSERVATION OF NUMBER

Conservation of number means that a student knows that changing the objects into a different arrangement does not change the number of objects. If we have four pencils, then move the pencils around or pick one up or put two down, we still have four pencils. As long as we don't add any or remove any, moving them around does not change the total. This sort of number permanence is not trivial. It requires having a sense of quantity (that there is a number of things, not just an amorphous blob). If a child is asked how many objects there are on the desk, and they count to find out, and then you move the objects and ask again how many, the child needs to have all of the big ideas and strategies required to do the counting, realize the end number represents the amount, remember that number, and realize that moving the objects around doesn't change the amount.

Revan shows that he probably does not strongly grasp conservation of number yet, because after he counts the seven candies, when I put them in the box and ask how many are in the box, Revan pauses, then reaches into the box to recount the candies.

By changing the orientation, did we change the total number of candies? Without conservation of number, Revan isn't sure and needs to count again. This is probable but not certain because Revan may just be doing what he thinks he is supposed to do in this interview, and like all of these principles, it is best observed in more than one setting.

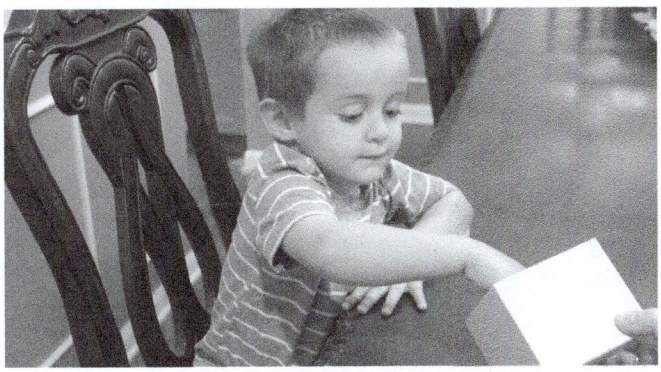

Harper shows that she is still working to build conservation of number because after she counts the seven cubes on the table correctly several times, when Kim puts the cubes back in the box, and asks how many are in the box, Harper says, "All of them," but has to count again to find the seven.

TRY IT

Help students build conservation of number by giving them counting tasks and sometimes rearranging the items to count. Then ask them how many for the rearranged group. If they are not sure or guess, ask if you have taken any out or added any. When they find that the amounts are the same by counting, comment, "So we had seven blocks. I put them in the box (rearranged) and there's still seven blocks. Interesting."

NEED FOR ORGANIZATION AND KEEPING TRACK

A student who moves the cubes as they count them to keep track of which have been counted feels the *need for organization and keeping track* (Fosnot & Dolk, 2001). This can be seen as Delila slid the cubes one by one to the side.

Students who do not yet feel the need for organization and keeping track often double count or miss objects. This double counting can be different than a student who does not yet have synchrony (one word for every object). A student may have synchrony in their counting but still inadvertently count some objects twice because they have lost track of which objects they've already counted.

Harper did not yet feel the need for organization and keeping track when she tagged each of the cubes, but because there was no movement, she lost track and double counted two cubes to report eight cubes while there were actually only six cubes on the paper.

TRY IT

When counting with children, visually organize your count, keeping track by moving objects to the side one at a time, or putting them in a box one at a time, or setting them apart in some way one at a time. Ask students who are showing that they feel the need for organization and keeping track to show how they are counting. Comment out loud how that seems to help them keep track.

COUNTING BACKWARD

Just because students can repeat the counting sequence forward does not mean that they can count backward. Counting backward can be helpful for students to learn the number sequence better and also lay the groundwork for hierarchical inclusion and later subtraction.

When I asked Savannah to count down, I started with "5, 4, . . . " and paused. She said, "5, 4, 3, 2" and stopped. I asked if she could do that from 10, "10, ..." She said, "10, 8, 9," and paused. I suggested, "7, . . ." and she said, "7, 6, 8." We tried again from 5, and she nailed it. Savannah could count backward from 5 but not from 10 *yet*.

When I asked Emmett to count backward, I suggested, "5, 4, . . . " and he started from 1, "1, 2, 3, 4, 5." I replied, "Very good counting! Can you go backward? Like 5, 4, . . . " and he smiled and said, "1, 2, 3, 4, 5." I asked, "What if we started at 10?" and Emmett gasped excitedly. I said, "10, . . . " and he put up both hands, counted down correctly all the way, synchronized with his fingers, and ended with "Go!"

> **TIP**
>
> Make the count *count*! That means making the count matter. For example, have students count the number of pencils they need for their group. If they miscount, they'll feel it because their group won't have enough or will have too many pencils. Have students count the number of snacks or papers or bouncy balls. Then hand those out. Do they have enough? Better count again! Creating spaces where students can reflect on and correct their counting helps them realize that the way they count has consequences. They need experiences where they can realize, "Oh, counting isn't just saying counting words and touching things; it actually represents the amount, and that matters!"

> ### TRY IT
>
> Count back with students! Start at a random number, not just 5 or 10, and choral count backward with them. When you have short bits of time during your day, make it a game to count back. Keep it about thinking, not speed. Go slow enough that students can think forward to help them if needed. And, when you reach the end, instead of "blast off" or "go," stop at 0.

> ### FREQUENTLY ASKED QUESTIONS
>
> **Q:** But Pam, my students can count forward, so of course they can count backward. Surely we don't actually have to work on counting down?
>
> **A:** Counting back might be harder than you imagine. You know the alphabet in order, right? A, B, C, and so on. Does that mean you can say the alphabet backward? Try it—start at M and say the alphabet backward. Do you find yourself saying some letters forward in order to then say them backward? Children might need experience doing that with the number sequence as well.

UNITIZING

When we can conceptualize many things as one group, we are *unitizing*. Unitizing means that you can consider simultaneously the objects in the set as units and the set as a unit (Fosnot & Dolk, 2001). Mentally coordinating these levels of units is cognitively difficult (Hackenberg et al., 2016, pp. 22–24). It takes experience grappling with situations where one needs to consider both levels.

> **TIP**
>
> If you have students who know more about the number system and want to keep counting back beyond 0 to –1, –2, –3, and so on, refrain from telling them they're wrong. Smile and just know that from now on you'll have to give them the ending point of 0, as in, "Today let's count down (backward) from 14 to 0!"

For example, when students begin to realize they can represent a quantity with one number, that three blocks (1, 2, 3) can be 3: one sound "three," one curvy symbol 3. This is unitizing: one sound, one symbol but three things.

Number of Objects	One Word	One Symbol
✋ ☐☐☐	Three	3

Unitizing also includes realizing that some amounts can be represented by one symbol (four objects by the numeral 4), but some amounts are represented by two symbols (ten objects by the two numerals 1 and 0 for 10). My young daughter, Abby (Figure 2.3), was playing school one day and asked, "Mommy, is twelve a one number or a two number?"

FIGURE 2.3 • Pam (author) With First-Grade Son, Craig, and Four-Year-Old Daughter, Abby

She was grappling with this bit of unitizing. Twelve, one sound (syllable), twelve objects, two symbols 12.

Number of Objects	One Word	Two Symbols
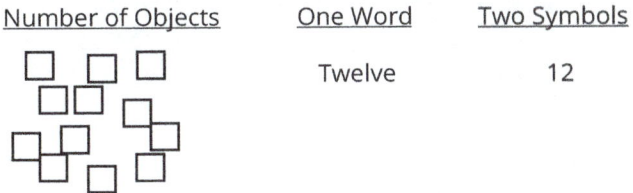	Twelve	12

Notice that while most of the single digits in English have names with one syllable, eleven has three syllables. Eleven objects is simultaneously 10 and 1 objects, three syllables, and two symbols, 11. How confusing for young learners!

Unitizing will come up time and time again at important sophistication junctures. Students will be asked to consider a new layer, where a new set of objects needs to be considered simultaneously as a set of individual objects and as one unit. In fact, one of the defining characteristics of each move to a new domain in the development of mathematical reasoning is a new expectation of unitizing, coordinating new and more levels of units. For example, multiplication requires students to consider the number in a group, the number of groups, and the total simultaneously. That's heavy-duty unitizing!

HOW TO DEVELOP COUNTING

Who knew there was so much rich detail in learning to count?! You might be wondering how to go about helping students develop all of the principles, strategies, and concepts you've just read about.

Watch this problem string to see how Kim represents 6-*ness*.

https://qrs.ly/3vgl255

One way is through Problem Strings. In Chapter 1, we saw how Melisa interacted with her students to help them build relationships that lead to an important Additive Reasoning strategy, Add a Friendly Number Over. We can use Problem Strings to support students as they make sense of counting as well! Working with kindergarteners, Kim Montague used the context of a bunkbed at a sleepover to help students reason about 6-*ness*. This is based on a New Perspectives on Learning task from Bunk Beds and Apple Boxes (Fosnot, 2025). She painted a vivid picture of six friends climbing up and down the bunkbed and what it looked like when she peeked in. She used the number rack as a model to represent how students are thinking about the total number of kids on the bunkbed, moving beads in and out of view on each rack to represent kids on the top or bottom bunk.

Students gave a private signal when they were ready to share their thinking, and Kim represented their thinking by moving beads and with equations. Through this back and forth between students and teacher, these students tinkered with all the different ways to make six, reasoning about decomposition within this accessible and engaging context of bunkbeds.

> ### FREQUENTLY ASKED QUESTIONS
>
>
>
> **Q:** What is the rekenrek (number rack) that Kim is using? Do I need one in my classroom?
>
> **A:** A rekenrek, or number rack, is a fantastic model to have in a K–2 classroom. You'll learn all about this fantastic model in Chapter 9.

THE NUMBER SEQUENCE IN THE TEENS

Many of the three-, four-, and five-year-olds that we interviewed owned several of these counting big ideas and used the strategies, but almost every one of them messed up the teen counting sequence, even those who got the sequence correct in the twenties and thirties and beyond.

I will use the word *teen* to describe the numbers 11 to 19, even though the words for eleven and twelve do not have the suffix teen. When I say teen, I mean those numbers that look like a 10 with a single digit taking the zero's place. In many languages, all of the numbers 11 to 19 follow a similar pattern.

When interviewing Revan, I asked him to count a bunch of Omnifix™ cubes. He began counting as he split apart the cubes one by one. He used the correct counting sequence through 11 and then kept repeating the same incorrect sequence 11, 14, 17, 18, 19.

Revan's counting:

1, 2, 3, 4, 5, 6, 7, 8, 9, 10,

11, 14, 17, 18, 19,

11, 14, 17, 18, 19,

11, 14, 17, 18, 19,

11, 14, 17, 18, 14, 17

He stopped when he ran out of cubes.

Emmett counted correctly to 14 on his fingers, then 16, 17, 18, 19, (pause) 21, 22, 23, 24, 26, 28, 29, 14, 28, 29, 14, 15, 16, 17, 18. Emmett might have kept looping back if I didn't smile at him, and he stopped. Savannah counted out loud with any objects correctly to 13, then 18, 19. Savannah stopped when she got to 19. When I asked if she could go any higher, she started over at 1.

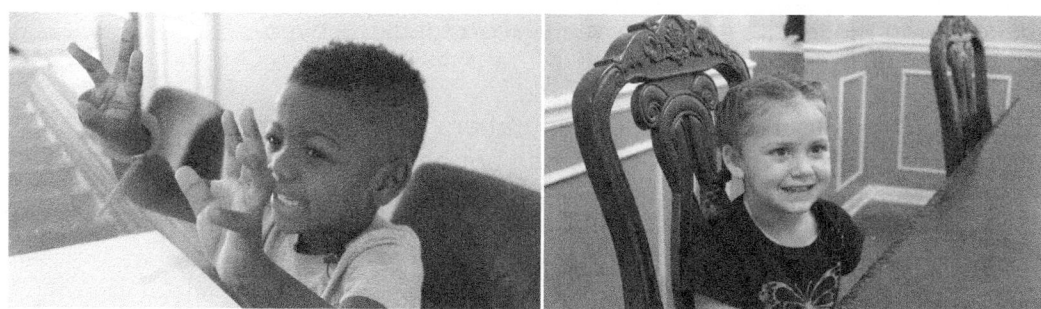

Many learners, like these, get the 1 to 10 sequence correct, but then struggle in the teens.

What is so hard about the teen sequence?

Much of the difficulty comes from two aspects: unitizing and language. The first is logical-mathematical knowledge, the second social.

As mentioned previously about unitizing, to understand the teen numbers, we need to make sense that a quantity can be represented by two symbols. Many students run into trouble as they are still constructing the idea that a quantity can be represented by one symbol (from 1 to 9), but are then asked to deal with two symbols for 11 to 19 (and beyond, but we'll get to that later). From a young learner's perspective, this could seem quite arbitrary.

Also, the English language does not make very clear the relationship between 1 to 9 and 11 to 19. Some of the teens, like *four*-teen, *six*-teen, *seven*-teen, *eight*-teen, and *nine*-teen clearly use *four, six, seven, eight, nine* sequence with the suffix -teen added on. But this is obscured by the fact that eleven, twelve, thirteen, and fifteen do not sound like numbers students have heard before. Eleven and twelve sound like completely new numbers, not related to one and two at all. Thirteen and fifteen introduce the sounds "thir" and "fif." What in the world is a "thir"?

Can you imagine how much easier it would be if the word patterns followed a more discernible pattern, like one-teen (11), two-teen (12), three-teen (13), four-teen (14 oh, it is!), and five-teen (15)?

Or, even better yet, what if the names of the numbers were even more congruent with the number system, like calling 11 "ten-one," 12 "ten-two," 13 "ten-three," and so on. The words would match the numbers!

We can help students make more sense of the teens by helping them realize these things for each teen. Let's look at 12 as an example:

- Quantity/magnitude: Know the number 12 is 10 and some extra, that 12 is made up of 10 and 2
- Name: Learn the name we've given that number, twelve
- Numerals: Recognize and use the symbols that represent 10 and 2, the number twelve, "12"

> ### TRY IT
> As students are counting into the teens, help them understand the connection between the numerals, the actual meaning, and the name: that 11 is really ten-one, and we call it eleven, 12 is actually ten-two, and we call it twelve, and so on.

You might think this is all silly. Just have kids memorize the new sequence of the teens, and be done with it. But did you know that many other languages actually follow the pattern? They do!

For example, in Russian you add надцать ("nadtsat"), which means "on ten" to the numbers 1 to 9. In Mandarin, you start with the word for ten and add on the numbers 1 to 9 (see Table 2.1).

TABLE 2.1 • Russian and Mandarin Numbers

NUMBER	RUSSIAN		MANDARIN	
11	одиннадцать	odin*nadtsat*	十一	shí yī
12	двенадцать	dve*nadtsat*	十二	shí èr
13	тринадцать	tri*nadtsat*	十三	shí sān
14	четырнадцать	chetyr*nadtsat*	十四	shí sì
15	пятнадцать	pyat*nadtsat*	十五	shí wǔ
16	шестнадцать	shest*nadtsat*	十六	shí liù
17	семнадцать	sem*nadtsat*	十七	shí qī
18	восемнадцать	vosem*nadtsat*	十八	shí bā
19	девятнадцать	devyat*nadtsat*	十九	shí jiǔ

This consistent pattern is super helpful. Let's help our students realize that even though this pattern is not consistent in English words, it is consistent in the actual number magnitudes. The teen numbers are 10 and a single digit.

FREQUENTLY ASKED QUESTIONS

Q: Why doesn't the English language follow a pattern in the teens?

A: Those number names, just like all language, developed over time. Names are social knowledge and often do not follow the pattern we wish they did. Help students realize that they can understand what the numbers mean and that the names are what we call them.

THE NUMBER SEQUENCE AFTER THE TEENS

Part of the brilliance of our base-ten place-value system is the use of only 10 digits – 0, 1, 2, 3, 4, 5, 6, 7, 8, 9 – to form any real number.

Our base-ten place value system is really that amazing.

Once children learn that sequence of numbers, they can use it to reason about the teen numbers and beyond. They can reason that in the twenties the numbers count up the same way, just with the additional twenty part, in the thirties with the additional thirty part, and so on.

Each larger set of 10 numbers repeats, 40 through 49, 50 through 59, 70 through 79, and so on.

We see this recognition in students when they are counting and they get to the bridge of a decade: 28, 29, . . . and they pause . . . and if you supply the 30, they are off again, 31, 32, 33, 34, 35, 36, 37, 38, 39. . . . At every decade, they can continue counting if they can just determine which decade they are in (at least until they get to 100 and then new landmarks appear; Fosnot & Dolk, 2001).

TRY IT

Use 100s charts to help students identify, analyze, and use the number sequence as they count. By starting at 1 and wrapping around after 9, the vertical columns show the repeating pattern in the ones digits and in the tens digits.

1	2	3	4	5	6	7	8	9	10
11	12	13	14	15	16	17	18	19	20
21	22	23	24	25	26	27	28	29	30
31	32	33	34	35	36	37	38	39	40
41	42	43	44	45	46	47	48	49	50
51	52	53	54	55	56	57	58	59	60
61	62	63	64	65	66	67	68	69	70
71	72	73	74	75	76	77	78	79	80
81	82	83	84	85	86	87	88	89	90
91	92	93	94	95	96	97	98	99	100

MEANING OF DECADES

Just as the teen numbers are related to the number sequence, the decades (multiples of 10) are similarly related.

The digits 1, 2, 3, and 4 are connected to 10, 20, 30, and 40. We could think about the numbers 1 to 9 as describing that many ones. In other words, 7 describes 7 ones. Then 70 describes 7 tens. Remember that unitizing means considering 70 simultaneously as 70 ones and 7 tens.

TRY IT

To learn the decades, use 100s charts to help students identify, analyze, and use the number sequence. By starting at 1 and wrapping around after 9, the vertical columns show the number sequence pattern in the tens digits. The decades are that tens digits number of tens. For example, 40 is 4 tens. The counting sequence shows up in the number of 10s.

1	2	3	4	5	6	7	8	9	10
11	12	13	14	15	16	17	18	19	20
21	22	23	24	25	26	27	28	29	30
31	32	33	34	35	36	37	38	39	40
41	42	43	44	45	46	47	48	49	50
51	52	53	54	55	56	57	58	59	60
61	62	63	64	65	66	67	68	69	70
71	72	73	74	75	76	77	78	79	80
81	82	83	84	85	86	87	88	89	90
91	92	93	94	95	96	97	98	99	100

Interview with Hiram.

https://qrs.ly/7dgl258

STUDENT INTERVIEW

You can watch an interview with Hiram, where you can see examples of a student owning, building, and approaching landmarks still on his horizon at the QR code on this page.

Conclusion

Counting is more than singing the song of the number names. To count well, children must learn several interrelated, nuanced big ideas and strategies. The good news is that understanding these important underlying big ideas and strategies and how to teach them means we can strategically teach counting. We can help students realize that counting is figure-out-able!

Discussion Questions

1. How are synchrony and one-to-one tagging related? How might you recognize if a student is using these in their counting?

2. How might you test for cardinality?

3. How might you test for conservation of number?

4. How might you test for hierarchical inclusion?

5. How do you think that these tests can simultaneously test and help develop these important counting big ideas?

6. What does it mean to you to make the count count? What's one or two ways that you can start making the count matter to your students?

CHAPTER 3

Counting Strategies

FIGURE 3.1 ● The First Level of Sophistication in Mathematical Reasoning

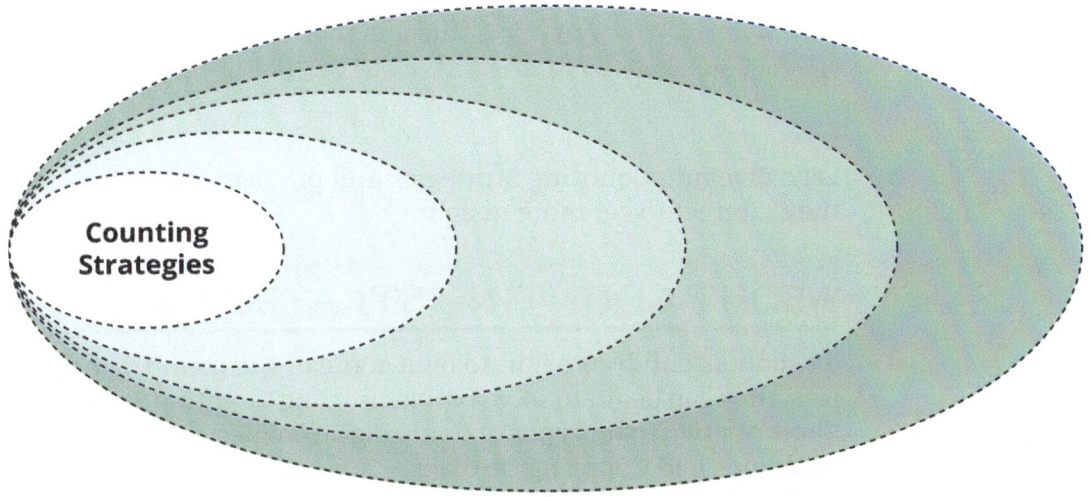

Source: Adapted from Math Is Figure-Out-Able at https://www.mathisfigureoutable.com/ with CC Attribution-NoDerivatives 4.0 International License.

It was early in the school year, and I had visited a first-grade class a few times. Students had been successfully solving word problems with numbers under ten, adding and subtracting.

This day, I give students another "addition" problem: "Marco has 7 crayons. Some are blue, and some are red. How many blue and red crayons could Marco have?"

Just as in the past few days, students begin working. I'm surprised to see that Daniel, who usually jumps right in, is just sitting there. I kneel down next to him.

"What are you thinking?" I ask.

He looks up at me, pauses, and says, "I don't know."

"Do you understand the question?"

"I don't know . . . I don't know what to do," he says and looks up at me.

I hand him two colors of crayons and repeat the question. He takes the crayons and counts 7 of them. Then looks at me.

I look around the room and notice that many of the students who are usually diving in to move counters on their desks, put up their fingers, or draw pictures on their papers are instead looking puzzled. Other students have found an answer and are moving on.

Something about this problem has *some* students stymied.

I was stymied. Why were some students able to do the previous problems but not now?

Let's dive into Counting Strategies and problem types and how they interact to get more insight.

ABOUT COUNTING STRATEGIES

As soon as children begin to own some of the counting principles, they can begin to solve problems using Counting Strategies. There are two main types of Counting Strategies: Early Counting strategies and Counting On/Back.

EARLY COUNTING STRATEGIES

When students begin to solve problems using a Counting Strategy, they often use the Counting Three Times strategy (Carpenter, 2014; Fosnot & Dolk, 2001). For addition problems, this often looks like counting out the first set, the second set, putting the sets together and counting again.

For example, solving 5 + 3 looks like counting out 5 objects (fingers, tally marks, and so on), counting out 3 objects, putting the sets together and counting all 8 objects. It doesn't matter what the model is. If the student is counting each set one by one, the strategy is Counting Three Times.

5 + 3 Using the Counting Three Times Strategy

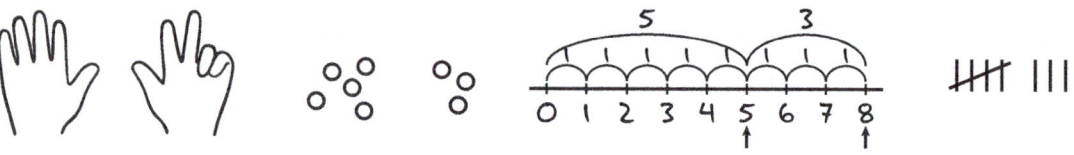

When using a Counting Three Times strategy, solving 9 – 5 looks like counting out 9 objects (fingers, beads), removing 5 objects one at a time (counting 5), then counting the remaining 4 objects. It doesn't matter what the model is. If the student is counting each set one by one, the strategy is Counting Three Times.

When students have to use "items to serve as markers for the count," counting out each set and the combination, Integrow calls this perceptual counting (MacCarty et al., in press).

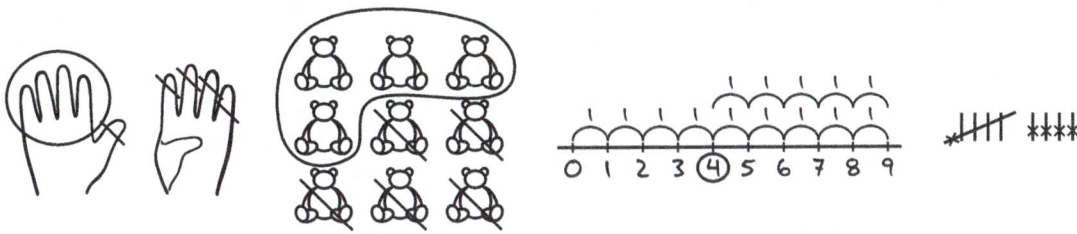

Students can also employ this strategy for simple multiplication and division problems.

TRY IT

Give students a problem involving groups of objects where the total is unknown, like "You and your two friends each get books. How many total books do we need?" Have counters and pencil/paper handy. If students count three people, count 4 for each person, and then count the total, the student is counting three times when multiplying.

Watch the clips to see how students might solve multiplication and division problems using a Counting Three Times strategy.

https://qrs.ly/mlgl25a

FREQUENTLY ASKED QUESTIONS

Q: Are you suggesting we give multiplication and division problems to young students who are still solving by counting one by one?

A: Yes! These tasks can help students reason about what is happening and use counting to solve the problems. As young students act out the problems and count one by one, they get better at the counting sequence and all of the counting principles from Chapter 2. Keep the numbers low and the situations familiar and figure-out-able. Refrain from telling students what to do to solve them. Encourage students to figure out what is happening and use what they know to find the answers. As students solve these problems, look for opportunities to nudge them to count on or back or even to use Additive Reasoning if they are ready.

A next Early Counting Strategy, which Integrow calls "Figurative" counting,

> involves coordinating the count (i.e., keeping track) in the absence of sensory materials. Knowledge with quantitative patterns, such as spatial patterns and finger patterns, support mental imagery for keeping track of the counts. When adding two quantities, the student counts from one for the first addend and then continues the count (while keeping track of the number of counts) for the second addend. (MacCarty et al., in press, p. 7)

When I asked Delila 8 + 5, Delila says, "We have 8," and then she puts up 5 fingers and then 3 fingers. "Then we add 1, 2, 3, 4, 5 more and then that makes 12, I mean 13."

When asked if it was 12 or 13 and how she knew, Delila pointed once to the 8 cubes and said, "8," then counted by ones: "9, 10, 11, 12, 13."

 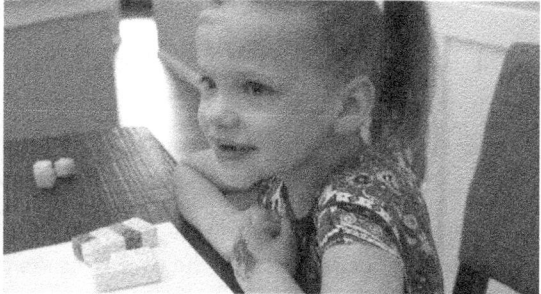

We can see that Delila is working to build the Counting On strategy because she begins by creating the first number but then sometimes doesn't actually use it. In the example of the cubes, she created both the 8-cubes group and the 5-cube group, but then she just points to the 8 and starts, "8. 9, 10, 11, 12, 13." She is using figurative counting working toward the Counting On strategy.

THE COUNTING ON, COUNTING BACK STRATEGY

When students do not have to count to create an addend in an addition problem but are able to start with that addend and then count on the rest one by one, they are Counting On. When students do not have to create the total in a subtraction problem but are able to start with that total and then count back one by one, they are Counting Back.

Watch a student solve a problem using this figurative counting strategy. Thank you to Integrow for the use of this video and help in understanding the difference between perceptual and figurative counting.

qrs.ly/s5gr0zc

Counting On and Back are more cognitively challenging for students for at least two reasons:

1. Students have to be able to conceive of that starting number without having to count up to it or create it. In 8 + 5, the student has to know 8-*ness* without having to count up to 8.
2. Students then have to keep track of the count and how many they are counting so that they know when to stop. For 8 + 5, the student keeps track of 9 and 1, 10 and 2, 11 and 3, 12 and 4, 13 and 5. They've reached 5 so they stop at 13. Keeping track of all of this simultaneously is cognitively difficult and represents more sophisticated thinking.

To Count On
8 + 5

1. Conceptualize 8.
2. Say the counting sequence while keeping the goal of adding 5 in mind.

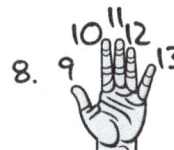

3. Stop because 5 fingers are up.

When Kaile was asked, "What is 8 + 5?" she looked forward and said quietly, "9, 10, 11, 12, 13," as she tapped 5 fingers on the table. She did not need to reproduce 8. She counted on from 8 one by one.

Watch students solve problems using the Counting On and Counting Back strategies.

https://qrs.ly/5egl25f

Once students are solving problems using Counting Strategies, let's build Additive Reasoning!

> ### TRY IT
>
> Give students the same kind of combining and separating problems as in the Early Counting Strategies section. Watch what students do. If they start with the first addend and count from there by ones, they are Counting On. If they start with the first number of the subtraction problem and count back by ones, they are Counting Back.

> ### FREQUENTLY ASKED QUESTIONS
>
> **Q:** Why do students clearly use Counting On/Back for some addition and subtraction problems but then revert to Counting Three Times or Figurative counting or guess and check for other addition and subtraction problems? What is happening?
>
> **A:** It's less important to have one description for a student's Counting Strategies and more important to help the student transition to Additive Reasoning strategies, thinking in larger chunks (see Chapter 4). If a student uses a less sophisticated strategy to solve a problem within a certain context, the chances are high that this is happening for one of two reasons: either the part of the problem that is unknown has changed or the problem type has changed, even though the problem involves the same number. We'll dive into that next. Read on!

PROBLEM TYPES

Math education researchers Thomas Carpenter and his colleagues produced a seminal work based on their research, *Children's Mathematics: Cognitively Guided Instruction* (2014). They observed, analyzed, sorted, and categorized children's strategies.

One of their major contributions is explaining the impact that different problem types have on students' ability to solve problems. They help us understand that some problems are easy for most children and some are much more difficult even though they seem like the same type of problem. What is it about the problems or the students that makes the difference?

In a nutshell, there are different problem types. These problem types and the location of the unknown in each problem have a huge impact on how students solve the problems and how easy or difficult they are for learners to solve.

First, I'll describe the major types of problems for young learners that have traditionally only been labeled as addition and subtraction: combining, separating, part–part–whole, and comparison.

Combining problems involve the action of bringing sets together. These might use words and actions like joining, adding, coming together, picking up, connecting, attaching, and receiving.

Separating problems involve the action of taking sets apart. These might use words and actions like leaving, dropping, popping (balloons), running/walking away, giving, removing, and taking away.

Part–part–whole questions do not involve action. They often involve different attributes of objects in a set like color (red and blue blocks), gender (girls and boys), size (large and small balloons), and species (cats and dogs for pets). In these problems, no action is happening. The problem is about the partitioned parts of a whole.

Comparison problems also do not involve action. Rather, these problems involve correlating amounts to determine how they relate to each other. They are based on the underlying concepts of more or less. In these problems, students are dealing with how much more or less.

FREQUENTLY ASKED QUESTIONS

Q: This sounds like abstract ideas that won't actually impact my teaching. Can I skip this part?

A: I recommend that you do not skip this part. Understanding how different problem types impact students using different strategies will be helpful. You will understand why some students are struggling more than other students with the same problems. You'll recognize which types of problems to use at the right time to best support your students' progress.

For each of these problem types, we can change which part of the problem is unknown. As you read the example problems in Table 3.1, Table 3.2, Table 3.3, and Table 3.4, ask yourself which you think would be easier for students to solve, which might be harder, and which are about the same level of difficulty.

TABLE 3.1 • Combining Problems With Different Unknowns

UNKNOWN	COMBINING
Result unknown	5 puppies are playing in the yard. 3 puppies run into the yard to play. How many puppies are playing in the yard?
Change unknown	5 puppies are playing in the yard. Some puppies run into the yard to play. There are now 8 puppies playing in the yard. How many puppies ran into the yard?
Start unknown	Some puppies are playing in the yard. 3 puppies run into the yard to play. There are now 8 puppies playing in the yard. How many puppies started in the yard at first?

Source: istock.com/GlobalP

TRY IT

Create three combining problems each with a different unknown in the context of insects crawling on a leaf and the numbers 2, 5, and 7.

TABLE 3.2 ● Separating Problems With Different Unknowns

UNKNOWN	SEPARATING
Result unknown	9 balloons are floating in the sky. 3 balloons pop. How many balloons are still floating in the sky?
Change unknown	9 balloons are floating in the sky. Some balloons pop. Now there are 6 balloons floating in the sky. How many balloons popped?
Start unknown	Some balloons are floating in the sky. Three balloons pop. Now there are 6 balloons still floating in the sky. How many balloons started floating in the sky?

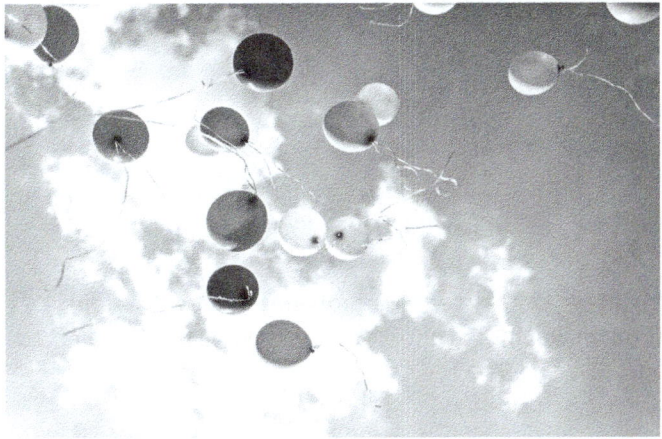

Source: istock.com/twpixels

TRY IT

Create three separating problems each with a different unknown in the context of eating fruit snacks and the numbers 3, 4, and 7.

TABLE 3.3 ● Part–Part–Whole Problems With Different Unknowns

UNKNOWN	PART–PART–WHOLE
Whole unknown	There are 6 large balloons and 3 small balloons at the party. How many total balloons are at the party?
Part unknown	There are 9 balloons at the party. Some balloons are small, and 6 are large. How many balloons are small?
Parts unknown	There are 9 balloons at the party. Some balloons are small, and some are large. How many small balloons and large balloons are at the party?

Source: istock.com/Annika Gandelheid

> ### TRY IT
>
> Create three part–part–whole problems each with a different unknown in the context of colored pencils and the numbers 5, 4, and 9.

TABLE 3.4 • Comparison Problems With Different Unknowns

UNKNOWN	COMPARISON
Difference unknown	There are shapes drawn on the paper, 4 squares and 9 circles. How many more circles are there than squares?
Compare quantity unknown	There are 4 squares drawn on the paper. There are 5 more circles than squares drawn on the paper. How many circles are drawn on the paper?
Referent unknown	There are 9 squares drawn on the paper. There are 5 more squares than circles. How many circles are drawn on the paper?

> **TRY IT**
>
> Create three comparison problems each with a different unknown in the context of tulips and daisies in a garden and the numbers 3, 6, and 9.

> **FREQUENTLY ASKED QUESTIONS**
>
> **Q:** Is it important for students to distinguish between problem types?
>
> **A:** No. Not at all. Teachers need to differentiate between problem types so they can plan which types to use at the best times to promote students' progress and also to understand why students are struggling. Students need to focus on figuring out what the problem is asking and then using their best thinking to reason about the solution.

Now that we recognize that there are different types of problems, let's dig into why some problems are more difficult for many students than others.

There are three major considerations for how difficult each of these problems are for children:

- Whether there is action in the problem
- What is unknown
- The size of the numbers

PROBLEMS INVOLVING ACTION/NO ACTION

If the problem involves *action*, it is easier for students directly modeling (acting out) the problem, counting one by one. Students can solve these problems using a counting strategy—the student acts out the problem, using what they know about counting, and finds the result.

If a problem does *not* involve *action*, it is harder for a student who is acting out and counting to solve the problem because there is no action to perform. The student is trying to *do something*, and if there is nothing *to do*, they can be stymied.

However, this is different for students who have built some Additive Reasoning. Once students have developed Additive

Reasoning, action doesn't really matter. Since these children are using the relationships between the numbers, they no longer depend on acting out the problem. They figure out what the problem is asking and then use the relationships between the numbers to figure it out.

For example, consider this part–part–whole problem: *There are 9 balloons at the party. Some balloons are small, and 6 are large. How many balloons are small?* A student using a counting strategy will wonder what they are supposed to do. A student using Additive Reasoning may think about the relationship between 9 and 6 and subtract to find 3 balloons, *even though* the problem did not call for a subtracting action. If the teacher tells a student who is using a counting strategy to subtract or worse to "minus," the counter will wonder why. "There's no removing/minusing happening in the story—why should I minus?"

Students using Additive Reasoning may even use the relationships in a way that does not match the action in the problem. In fact, this happens often.

For example, consider the combining-change unknown problem: *5 puppies are playing in the yard. Some puppies run into the yard to play. There are now 8 puppies playing in the yard. How many puppies ran into the yard?* A student who is acting out and counting will act out the combining action of puppies running into the yard. For a student who is reasoning additively, they might think 8 subtract 5 is 3, so 3 puppies ran into the yard. Even though the problem is combining, the student using Additive Reasoning subtracts to find the answer.

TIP

When trying to determine if the problem is combining or separating or part–part–whole, distinguish the action in the problem from the action you are taking to solve the problem. Act out the problem, and describe the problem type based what is happening, not the way you or your students are solving it.

TRY IT

Consider this problem: *There were some marbles in the jar this morning. Denisha took out 3 marbles at lunch time. There are now 5 marbles in the jar. How many marbles were in the jar this morning?* How might a student who is using a counting strategy solve it differently than a student who is using Additive Reasoning?

WHAT IS UNKNOWN

The difficulty of problems also depends on what is *unknown*, especially for students who have not yet built Additive Reasoning.

When the result is unknown, either for combining or separating, a child who is acting out the problem and counting does just that. They might create the first set, the second set, do the action, and count the result, Counting Three Times. Or they start with the first number, do the action, and land on the result, Counting On/Back.

However, when the change or start is unknown, the problems become much more difficult. If a child is Counting Three Times, such problems are impossible to act out because they don't have a change or start value to use. They may try to guess and check and, with a lot of perseverance, have some luck. If a child is Counting On/Back, they may be able to solve change-unknown problems by beginning with the start and then Counting On or Back. But if the problem is start-unknown, that same child will have to guess and check a starting number.

Therefore, if a child does not own any Additive Reasoning, which means they are acting out and counting, often the child's only recourse is to guess and check. This can feel arbitrary and exhausting.

Part–part–whole problems do not involve action. These can be more difficult for students using Counting Strategies because there is nothing to do. We often see these students count out the two sets of numbers in the problem but then stop, not knowing *what to do* next.

Comparison problems are extremely difficult for many students using Counting Strategies. There is no action, and the situation of comparing is more cognitively demanding than combining, separating, or considering the part–part–whole situation. It can be difficult for young students to even decide what the question is asking. If students are using Additive Reasoning, once they figure out what the question is asking, they can use those additive relationships to solve.

FREQUENTLY ASKED QUESTIONS

Q: For comparison problems, you say that they are difficult for students who are using Counting Strategies. I just demonstrate for students that they can line up objects to compare them and count the difference. Won't that make it easier for them to solve?

A: Might it make it easier in the moment for a student to answer that question? Sure. But remember that the goal is to build

(Continued)

> (Continued)
>
> reasoning, not just get answers. Also, for many students, they won't realize when to do it. Do they always take the numbers in a word problem, line them up, and count the difference? When don't they? Many teachers start teaching key words when they find students getting confused about what to do when. Then students try to memorize those key words and what to do with each different key word. Instead, focus on the sense making, helping students realize what is actually happening in the problem while building Additive Reasoning. The key is to realize that Additive Reasoning is the goal.

Take heart that all the problems are easier for students who have built and are using Additive Reasoning. Our goal is to help students develop Additive Reasoning!

THE SIZE OF THE NUMBERS

The difficulty of problems also depends on the size, or magnitude, of the numbers involved. Students may Count On/Back when the numbers are small enough because they have enough experience to be able to conceptualize the starting number and count on from there. But that same student might get into the teen numbers and be less clear or confident about starting from that higher, less familiar number.

We see this happen with students who have built Additive Reasoning with numbers within 10, reasoning confidently with those numbers, who then revert to Counting On/Back in the teens and beyond.

In fact, this happens at each major place-value juncture. Students revert to a less sophisticated, but well-established strategy when the numbers get bigger or more complicated. This usually only lasts for a short time if you work to help students gain the experience with the bigger numbers and keep developing the more sophisticated strategies.

DEVELOPING COUNTING STRATEGIES

In the first part of this chapter, you learned about Counting Strategies. In this section you'll learn more about developing these strategies.

DEVELOPING THE EARLY COUNTING STRATEGIES

Develop the Early Counting Strategies by giving students contextual action-result unknown problems to solve. By acting out and counting, the students will naturally count three times: count the first set, count the second set, perform the action, and count the third set. Help students continue to develop all the counting foundations you learned in Chapter 2, Counting Three Times, and Figurative counting by doing the following:

- Give students appropriate problems to solve by starting with action-result unknown.
- Make the context realizable (Fosnot & Dolk, 2001) by creating a discussion about the context with students so that they understand the situation.
- Act out the problems as needed so students understand what is happening in the problem.
- Circulate and work with students, helping them learn to represent their thinking. See Chapter 9 for more on representing thinking with a model.
- Facilitate a Class Congress (discussion), eliciting, representing, and comparing students' strategies.
- Facilitate Problem Strings to help students learn to count better. See Chapter 8 for more on Problem Strings.

More details on early counting strategies.

https://qrs.ly/ylgl25g

FREQUENTLY ASKED QUESTIONS

Q: Is it important for students to Count Three Times before they start Counting On/Back?

A: No, if you have students who are already Counting On or Counting Back, you probably just missed when they were Counting Three Times. It may have been before they knew you, or it may have happened quickly. Either way, keep pressing forward.

DEVELOPING THE COUNTING ON/BACK STRATEGY

Developing the Counting On and Counting Back strategies involves continuing to give students problems to solve and, while students are solving them, helping students to realize that they don't have to create the first set, that they can start with that number and Count On or Back.

> **TIP**
>
> Once students begin to Count On, you can work to help them realize that it is more efficient to Count On from the larger. At the same time, you'll want to work on building Additive Reasoning. If students skip Counting On from the larger number and start using Additive Reasoning instead, that's great.

Use these specific classroom activities with your students to help them develop Counting On/Back.

- Counting variations
- Grouping/covering one set of the objects/numbers
- Give students new problem types to solve: part–part–whole and action-change unknown
- Eliciting, representing, and discussing the strategies of students who are Counting On/Back
- Problem Strings

Let's get into more detail for each of these teaching activities.

COUNTING VARIATIONS

Count aloud with students, but do not always start with 1. Say, "Let's count, but today we'll start with 4."

Sometimes count down from a random number. Sometimes just ask for the next number or the previous number.

> **TIP**
>
> Remember that this is not about speed. Ask students to briefly pause so that everyone gets a chance to think before they call out the next number. This sends the message that you value thinking, not just quick recall.

Vary the number you start with. Do this counting from different numbers often, in short bursts.

As you choose starting numbers, use numbers under 10 until most students can name the number before or after those numbers. Then move into the teens, then the twenties, and so on.

COVERING THE OBJECTS/NUMBERS

As your students are solving action-result unknown problems, circulate and listen to their strategies. When you think students might be ready because they are successfully using many of the counting big ideas and strategies from Chapter 2, and they are Counting Three Times, you can begin to gently suggest Figurative counting and then Counting On by covering the first set of objects and then counting on the second set with the students (Table 3.5).

TABLE 3.5 • Examining Strategies

COVERING		
FOR A PROBLEM LIKE:	**WHEN THE STUDENT...**	**THE TEACHER CAN...**
5 puppies are playing in the yard. 3 puppies run into the yard to play. How many puppies are playing in the yard?	counts out 5 blocks and 3 blocks and is starting to count all of the blocks	gently cover the 5 blocks and ask, "How many?" When the child says 5, the teacher can repeat, "Ah, so you know 5," and begin to slowly count the rest of the blocks *with the student*.

When you cover the set of objects for the first number, you can ask, "Do you know what one more is?" or "Do you know what comes after 6?" or "What is the number after 6?" If the student does not know, allow them to count to find out. Then gently reaffirm that the number does indeed come right after 6. "Ah, so you counted up and found that 7 comes after 6. Nice. So 6, 7, 8, 9..."

ELICITING AND REPRESENTING THE STRATEGIES OF STUDENTS WHO ARE COUNTING ON/BACK

As students are solving word problems, circulate and interact with students to find out how they are reasoning—are they using a Counting Strategy (which one?) or Additive Reasoning? When many or most of your students are using an Early Counting strategy and a few of your students are starting to Count On, find both a student who is Counting Three Times and a student who is Counting On, and ask each to share in the Class Congress. Represent their strategies. A sample final display is shown in Figure 3.2.

TIP

When a student is sharing, you can ask them to share just the part you're looking for. For example, if a student starts by explaining that they Counted Three Times and then decided to Count On, you can ask them to just share the Counting On strategy. Students do not have to walk through their whole process. They can share just the part you need for that discussion.

FIGURE 3.2 • Sample Final Display Showing Counting Three Times and Counting On

You have 8 stickers in your book. Your friend gives you 2 more stickers. How many stickers do you have now?

Counting 3 Times

```
 1  2  3  4  5
[1][2][3][4][5]
 6  7  8  9  10
[6][7][8][9][10]
```

Counting On

(8) 9, 10

10 stickers

TIP

As you look for students to share their strategy, always consider whose voice has not been heard for a while or who you could position today as a sense maker. Don't let the quick, louder students dominate the discussions.

Create a discussion around efficiency, asking and noticing the following:

- Which took longer?
- Does one seem like it would take less time?
- Mathematicians often try to do things efficiently, taking less effort.
- Which do you want your brain to do next time?

GIVE STUDENTS NEW PROBLEM TYPES TO SOLVE

When your students are successfully solving action result-unknown problems *well enough*, it's time to sprinkle in some part–part–whole whole unknown and then action change-unknown problems.

As students are solving these problems, encourage them to use their newly formed Counting On or Back strategies. They might have to guess and check a bit. Keep making it about sense-making. Keep doing all of the other activities so that your students will be building Additive Reasoning and can start using additive relationships instead of counting one by one.

PROBLEM STRINGS

You can use Problem Strings to help students begin to Count On and Count Back. The following Problem String suggests the same starting number (Table 3.6). If students are Counting Three Times to solve each problem in the string, let them do it, but don't represent that strategy. Ask students to share who are

Counting On, and represent that by writing the starting number and circling it, then counting by one from there. Notice aloud that the starting number is staying the same. After a couple of problems, wonder aloud if other students want to try Counting On like they are hearing their classmates try.

TIP

Refrain from backing up just because students struggle with the new problem type. Give students opportunities to grapple with these new ways of asking problems. The new problem types might actually help students move forward. Don't up the numbers at the same time though. Only change one thing at a time.

TABLE 3.6 • Problem String Using a Number Rack

Number Rack	Prompt
●●●●● —— ○○○○○ ●● —— ●●●○○○○○	"Let's focus on the left side. How many beads do you see?" *Model student thinking by moving beads and with expressions.*
●●●●● —— ○○○○○ ●●●● —— ●○○○○○	"How many beads here? Did anyone start with 5? Could you?"
●●●●● —— ○○○○○ ● —— ●●●●○○○○○	*Repeat.* "How many is 1 more than 5? What number comes after 5? How does that help?"
●●●●● —— ○○○○○ ●●● —— ●●○○○○○	"Did anyone start with 5 and then count 3 more?"

Sample final display for this Problem String on a number rack:

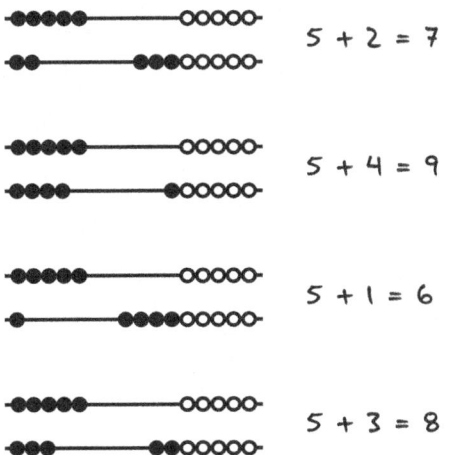

Download a handy PDF for this Problem String.

https://qrs.ly/obgl25h

More Counting On/Back Problem Strings.

https://qrs.ly/gegl25j

Sarah Hempel is a first-grade teacher, circulating and working with students who are solving the following problem:

> Jacob has 4 toy cars. He finds 3 new toy cars. How many toy cars does Jacob have now?

As Sarah kneels down next to Keaten, she notices that Keaten is drawing all of the toy cars, one by one. She is Counting Three Times.

Sarah has observed that Keaten has been successful Counting Three Times, understanding the problems, and correctly creating and counting the sets with counters. Keaten has also been counting with the class, starting at random numbers. Sarah decides to nudge Keaten toward Counting On.

Sarah asks, "Tell me what you're thinking."

Keaten replies, "These are the 4 cars. I'm drawing the 3 new cars."

Sarah gently puts her hand over the 4 cars and says, "So, you've got 4 cars here?" When Keaten nods, Sarah writes a large numeral 4 and asks, "What if you just had one more car? How many would you have?"

Keaten looks up and says, "5."

Sarah holds up a finger and repeats, "5. And one more car?"

Keaten says, "6."

Sarah holds up another finger, "6. And one more?" prompts Sarah.

When Keaten says, "7," Sarah holds up one more finger to have a total of 3 fingers up.

"Did we get enough? You drew 4 cars here. And then we got how many more?" she asks as she looks at Keaten.

Keaten looks at Sarah's fingers and says, "Yes, we did 3 more. So it's 7."

Sarah then shows Keaten how she could write that on her paper.

Before she moves on, Sarah nudges Keaten to think about what they just did together, "So it seems pretty helpful to start with the 4 cars and then add the 3 on, one car at a time? Nice work! And when your brain does that, you could show your thinking like this on your paper. What do you think?"

Sarah reports:

> To help students begin to Count On, I find that it's very helpful to ask students if they know just one more. And then one more. We count on together and then make sure we counted enough. Once students are comfortable with that, I show them how to represent that on their paper. They may need a few more times where I encourage them to Count On and help them represent their thinking, but as we talk about it as a class, they start to realize that it's so much more efficient. It's so great to see their minds literally growing right before me!

FREQUENTLY ASKED QUESTIONS

Q: My students are successfully Counting On, but they keep getting mired in Counting Back. Some students correctly count back to the answer, but some keep ending 1 too high or low. They've got the counting sequence down, they know what's happening in the problem, they just end on the wrong number. What should I do?

A: It is difficult to count backward! Keep working on helping students make sense of combining and separating problems, but also work on developing their Additive Reasoning. As soon as they are using additive relationships to solve problems, they will use these relationships to subtract. It's more important that they develop Additive Reasoning than they can solve problems counting one by one backward.

Conclusion

It is a beautiful thing to watch young children develop the underpinnings of counting and Counting Strategies. It is even more wonderful to actively promote that development through deliberate math-ing with tasks, questions, and instructional routines. When we help students learn that they can reason through problems, always building more sophisticated reasoning, students realize that math is actually figure-out-able!

Discussion Questions

1. How would you describe the difference between the Early Counting Strategies and the Counting On/Back strategy?
2. Create a combining change unknown problem.
3. Create a separating start unknown problem.
4. Create a part–part–whole part unknown problem.
5. How do problem types interact with Counting Strategies?
6. Which problem types are easiest for students who are using an Early Counting strategy (Counting Three Times or using Figurative counting)?
7. Which problem types are the hardest for students who are using an Early Counting Strategy (Counting Three Times or using Figurative counting)?
8. Which of these problems is probably harder for a student who is using an Early Counting strategy (Counting Three Times or using Figurative counting)? Why?
    a. 10 apples fell from the tree to the ground. I picked up 4 of them. How many apples are on the ground?
    b. Some apples fell from the tree to the ground. I picked up 4 of them. Now there are 6 apples on the ground. How many apples fell from the tree?
9. Which of the problem types should you hold off asking until students are beginning to use some Additive Reasoning? Why?
10. What are you taking away from this chapter? What are you wondering about?

PART III

Developing Additive Reasoning

Chapter 4: The Major Strategies for Addition Within 20

Chapter 5: The Major Strategies for Subtraction Within 20

Chapter 6: The Major Strategies for Double-Digit Addition

Chapter 7: The Major Strategies for Multi-Digit Subtraction

CHAPTER 4

The Major Strategies for Addition Within 20

FIGURE 4.1 • The Second Level of Sophistication in Mathematical Reasoning

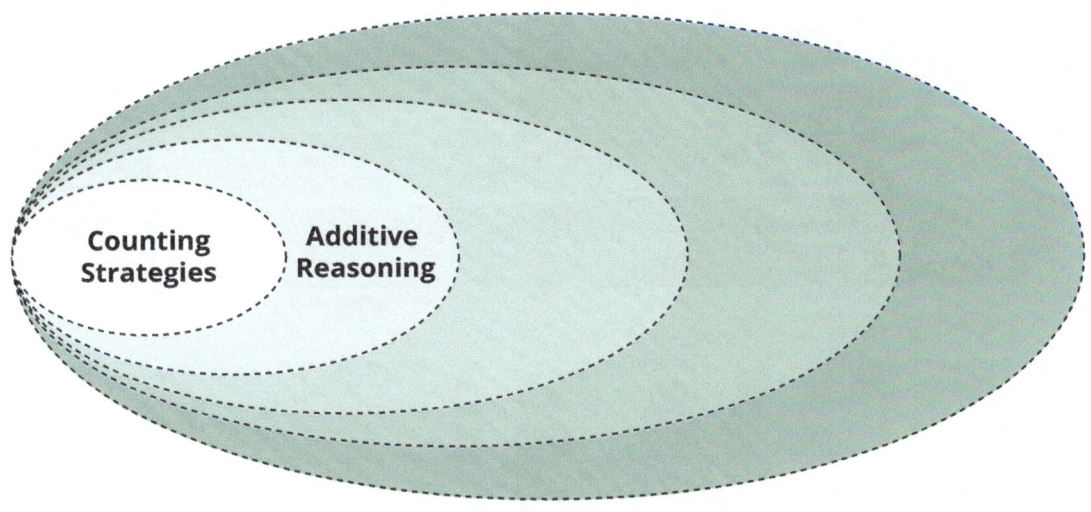

Source: Adapted from Math Is Figure-Out-Able at https://www.mathisfigureoutable.com/ with CC Attribution-NoDerivatives 4.0 International License.

Kim, a guest teacher in a first-grade class, begins the lesson: "Today we are going to use the math rack to solve problems." Kim describes that she is going to cover the math rack with a sheet of paper, move beads to the left, "flash" to show the beads for a couple of seconds, and then ask the students how many beads they saw.

Kim covers the rack, moves some beads, and then after about five seconds of revealing the rack, she covers it again and asks for thumbs-up for how many the students saw.

She calls on Nico, who says, "14."

Kim asks how Nico knows that.

Nico answers that there are 5 red on top, five red on bottom, 2 white on top, and 2 white on bottom. Kim asks students again for a thumbs-up, but this time if they agree with what Nico just said.

Kim calls on Kade, who was signaling he agrees. "What did he say? Did you hear him?"

Kade responds, "Um, kind of."

Kade appears nonplussed, so Kim says, "You can ask him if you're not sure what he said. Do you want to ask him? You can just say, 'Nico, tell me again what you saw.'"

After Kade turns and asks Nico to tell him again, Nico repeats his thinking, and Kim simultaneously represents it with an equation.

Kim then asks if anyone saw the problem in a different way, "You saw that there were 14, but you saw it in a different way?"

Kim calls on Sophia, who pauses for several seconds and then says, "10 and 4."

Kim repeats encouragingly, "You saw it as 10 and 4. Interesting! What's the 10, Sophia?"

Sophia smiles, and after a few seconds, Kim moves the 4 white beads to the right and asks, "Did you see these right here?"

As Sophia nods, Kim summarizes, "14 can be thought of as 5 + 5 + 2 + 2, or (*writing "= 10 + 4" on the board*) as 10 + 4."

She continues the equation: "How many were on top?" (*writes* 7) "And how many on the bottom?" (*writes "= 7 + 7" on the whiteboard*).

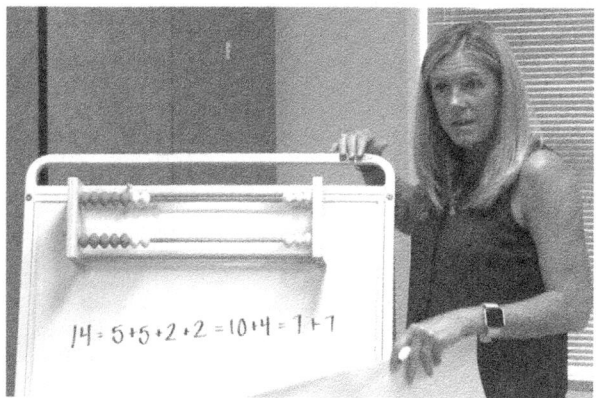

For the next problem, Kim repeats covering the rack, moves beads unseen, and uncovers to reveal 7 on the top and six on the bottom.

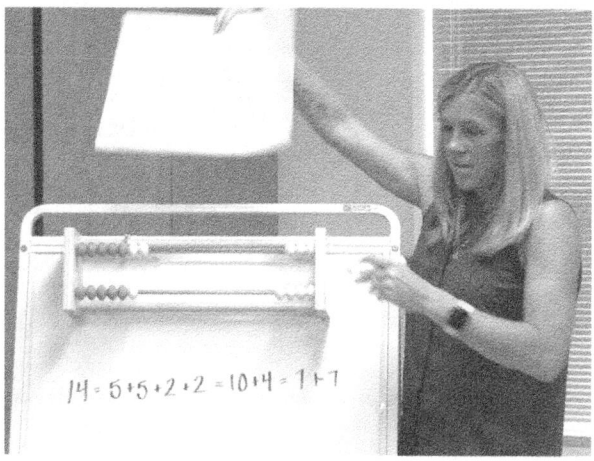

When Kim asks, "Sydney, what did you see?" Sydney answers, "13."

Kim responds, "You saw 13, how did you see those 13?"

As Sydney says, "'Cause I saw 10 red and 3 whites," Kim starts a new equation on the board, "13 = 10 + 3" and moves the three white beads to the right a bit to separate the 10 red and 3 white beads.

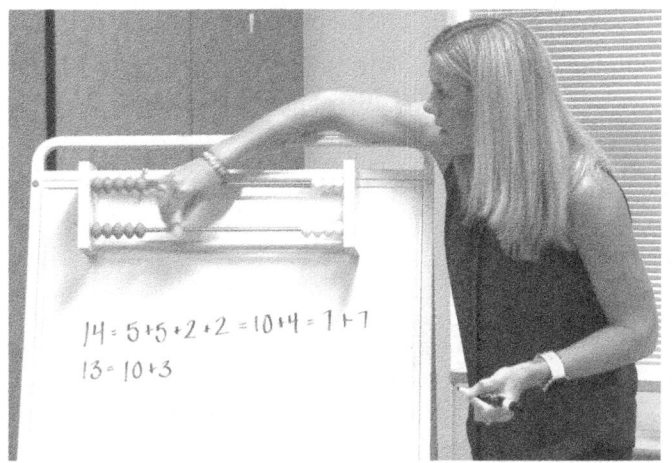

Kim asks, "Did anyone see it a different way? Brady, I thought I heard you start to say something. You said it's just..."

Brady repeats, "One less."

Kim asks, "One less than what?"

Brady responds, "14."

Acknowledging the connection to the previous problem, Kim moves beads and says, "We just had 7 and 7 (*pushing one bottom bead to the left*) which was 14. And you said (*she slides that bead back to the right*) it's just one less. So I could also say that 13 is the same (*continuing to add to the previous equation*) as 7 + 7 − 1.

Chapter 4 • The Major Strategies for Addition Within 20   73

Brady exclaims, "Whoa!"

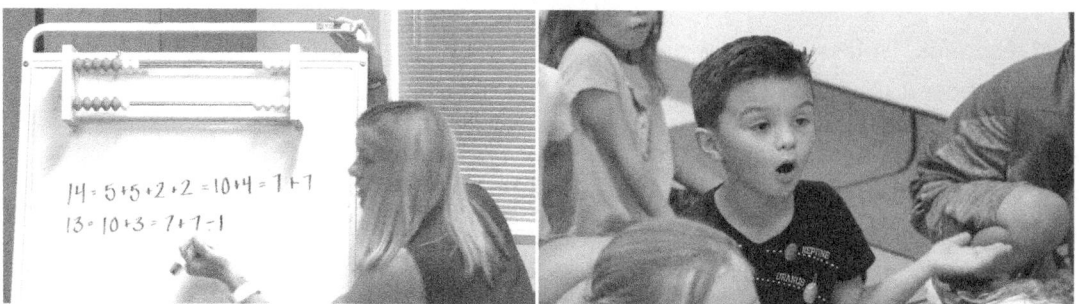

Kim finishes the problem: "We had 7 on top and 6 on bottom," adding to the equation so it reads "13 = 10 + 3 = 7 + 7 − 1 = 7 + 6."

Kim continues: "Really cool! Okay, ready for another one?" She covers the rack with paper "Here we go, ready?" Kim reveals 7 on top and 8 on bottom.

Two students excitedly call out, "It's just one more!"

Kim asks, "Luke, what did you see?"

When Luke responds, "15," Kim brings in others and asks about reasoning: "Did anyone else see 15? Lots of connections to 15. Luke, how did you see that 15?"

Luke answers, "Because this number was just one more."

Kim helps make sure everyone is following by asking Luke to specify, "One more than what?"

Luke answers, "14."

Kim says, "One more than 14. You saw 15, and you said (*writing out "15 = 7 + 7 + 1"*) earlier we had 7 and 7, and you said, well, it's just one more.... And how many are on top and bottom?"

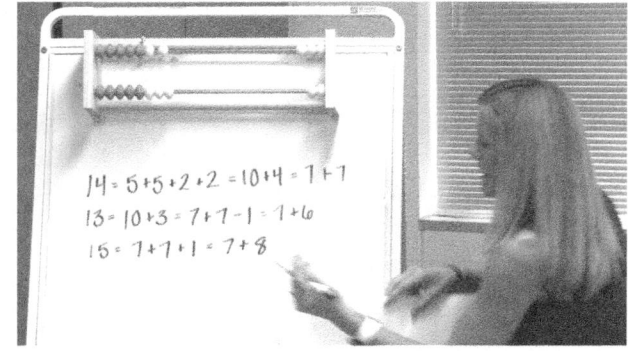

As students call out 7 and 8, Kim adds to the equation "15 = 7 + 7 + 1 = 7 + 8."

Kim next covers the rack and then reveals 8 on top and 8 on bottom.

When Gemma says that there are 16 beads and Kim asks how she saw that, Gemma replies, "Just one more than 15."

Kim says, while moving beads and writing equations, "You saw 16, and you said to yourself (*begins manipulating rack to model numbers as she says them*) earlier we had 7 and 8, so now do we just have 8 plus 8? (*writing "16 = 8 + 8"*) Who heard how Gemma just solved that? She said it's just one more. Just one more than what, Juliette?"

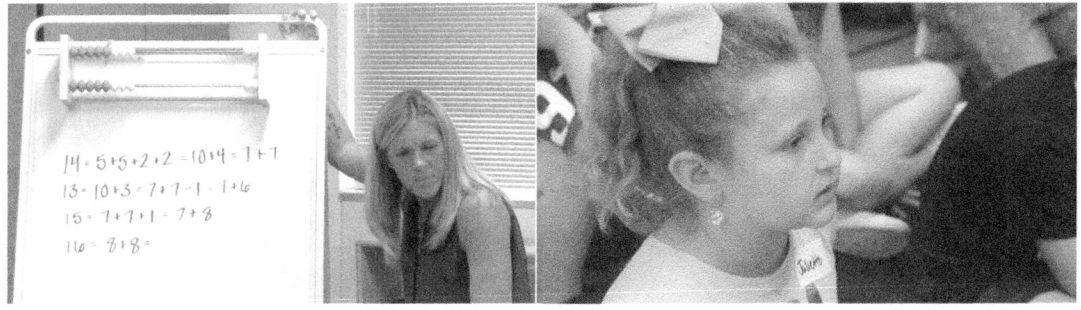

Juliette says, "One more than 15. Um, with one bead taken away from the bottom, and so it's just like 15, but when you add it, it's 16."

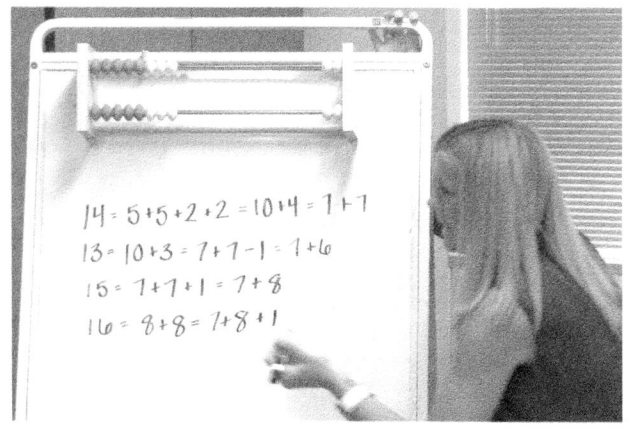

Kim revoices Juliette's thinking and makes it visible: "Earlier we had 7 plus 8 (*continuing the equation* "$8+8=7+8+1$") now we have 7 plus 8 plus 1 more."

For the last problem, Kim covers the rack with paper, moves beads, and reveals 8 on top, 9 on bottom.

Similar conversations follow, with Kim focusing on "How do you know?" moving the beads appropriately.

When Cash sees 8 + 9 as 10 + 7

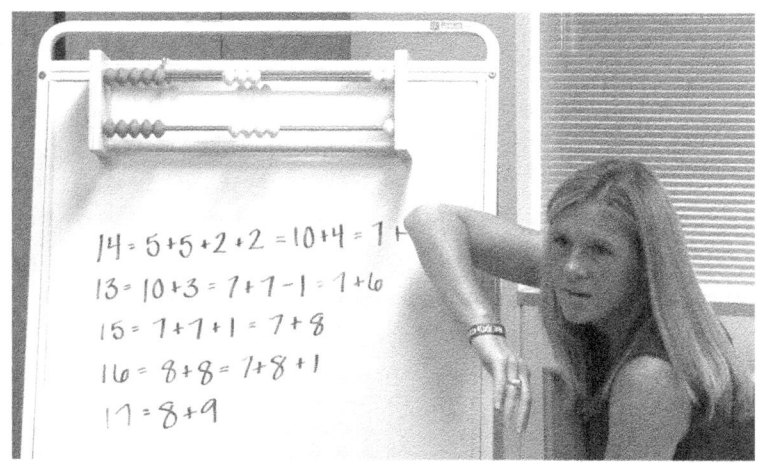

When Charlotte sees 8 + 9 as 8 + 8 + 1

Kim helps bring the major idea of *using doubles* into focus, moving beads and underlining numbers: "Check this out. We started off with 7 and 7 (*underlines 7 + 7*), and you said to solve 7 and 6, it's just one less (*underlines 7 + 7 – 1*), and then 15 is like 7 and 7 and 1 more (*underlines 7 + 7 + 1*). Then we did 8 and 8 and our friend just said that 17 is like 8 and 8 and 1 more" (*underlines 8 + 8 + 1*).

Kim smiles. "Ya'll are such good thinkers."

These students are learning to think in bigger chunks of numbers than one at a time. They are developing Additive Reasoning.

To watch this clip and see these students reasoning in action, see the QR code on this page.

Reasoning in action.

https://qrs.ly/9egl25m

FREQUENTLY ASKED QUESTIONS

Q: In Kim's Problem String about doubles, she covered the rekenrek so students saw the beads for only a short time. Why did she do that?

A: This is called a Quick Image, where you *flash* the image for a brief time. This is intended to help students create spatial, visual images in their minds. This can help students shift from counting one-by-one and instead start grouping the beads. They can use the 5- and 10-structure of the rack to help them think about the beads in groups.

ADDITIVE REASONING

FIGURE 4.2 • Additive Reasoning

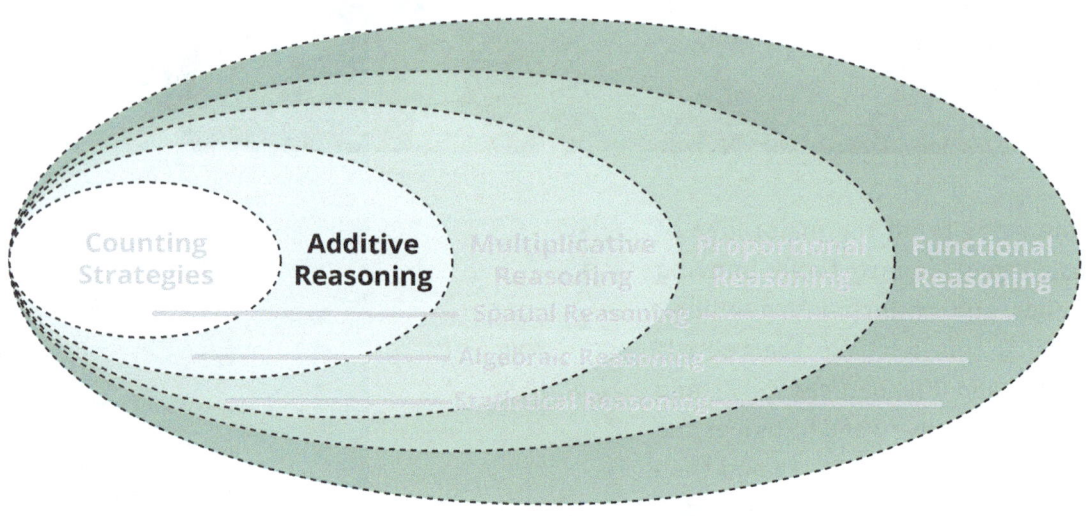

Source: Adapted from Math Is Figure-Out-Able at https://www.mathisfigureoutable.com/ with CC Attribution-NoDerivatives 4.0 International License

Additive Reasoning builds on Counting Strategies (see Figure 4.2). Reasoning additively means using what you know about numbers and the relationships between the numbers, and not just the counting sequence.

For instance, when finding the total of 98 and 55, students could reason additively that by subtracting 2 from 55, adding it to 98, and combining the remainder would give you the sum of 153.

Additive Reasoning is essential to develop, just like Counting Strategies, because it will continue to be integral to every subsequent domain of reasoning (Figure 4.3).

FIGURE 4.3 • What does Additive Reasoning look like?

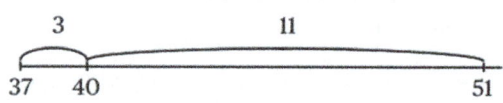

Counting Strategy	Additive Reasoning
In the example above, the student is counting one by one, shown by individual jumps on the open number line to solve the problem 37 + 14.	In this example, the student uses their knowledge of numbers in relation to units of 10 to solve the same problem, 37 + 14.

Source: Adapted from IES (2024).

78 • Part III • Developing Additive Reasoning

TIP

Developing Additive Reasoning does not mean a student will never count again. Counting the number of pencils you need or counting how many students are in line will still happen. However, that does *not* mean a student should be counting if they have access to a more sophisticated way of solving a given problem.

Chapters 4, 5, 6, and 7 will help you develop the major additive relationships that lead to students building and using the major additive strategies.

ADDITIVE STRATEGIES

Figure 4.4 shows a zoomed-in view of the Additive Reasoning domain with the major Additive Strategies shown.

A student solving an addition problem using a counting strategy.

https://qrs.ly/s2gl25n

FIGURE 4.4 • Sequence of Additive Strategies

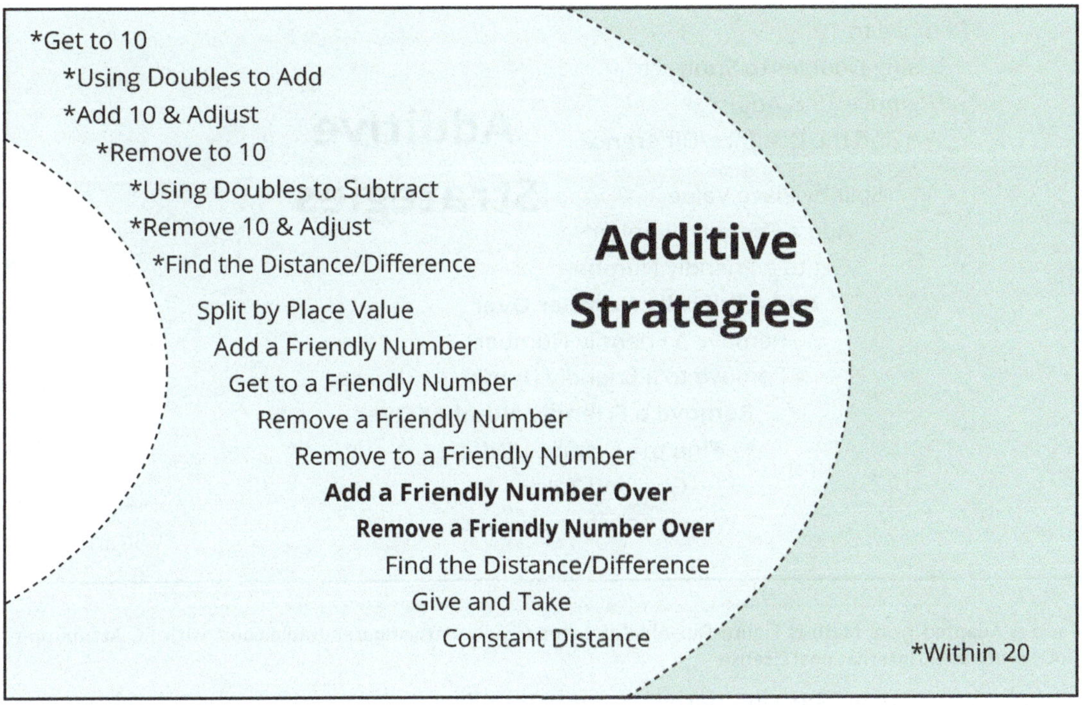

```
*Get to 10
    *Using Doubles to Add
   *Add 10 & Adjust
       *Remove to 10
        *Using Doubles to Subtract
       *Remove 10 & Adjust
        *Find the Distance/Difference
       Split by Place Value
       Add a Friendly Number
      Get to a Friendly Number
     Remove a Friendly Number
    Remove to a Friendly Number
     Add a Friendly Number Over
    Remove a Friendly Number Over
      Find the Distance/Difference
           Give and Take
           Constant Distance
```

Additive Strategies

*Within 20

Source: Adapted from Math Is Figure-Out-Able at https://www.mathisfigureoutable.com/ with CC Attribution-NoDerivatives 4.0 International License

Notice that the strategies for addition and subtraction within 20 are listed with an asterisk (*). The strategies are shown from left to right, top to bottom to indicate a general order in which you can help students develop the relationships that lead to these strategies. Some strategies are lined up, justified to the left—this means they can be taught around the same time and in either order because they are at about the same sophistication level.

A student solving an addition problem using an additive strategy.

https://qrs.ly/rogl25o

For example, *Using Doubles to Add could be taught before or after *Add 10 & Adjust. These two strategies are developing similar but different things so it doesn't matter the order.

There are two possible equivalent sequences for developing these Additive Strategies. The sequence in Figure 4.5 suggests that after students have developed two addition strategies (Add a Friendly Number and Get to a Friendly Number), and two subtraction strategies (Remove a Friendly Number and Remove to a Friendly Number) they can work on the two over strategies, Add a Friendly Number Over for addition and Remove a Friendly Number Over for subtraction. This means that the addition and subtraction strategies are a bit more mixed together.

FIGURE 4.5 • Alternate Sequence of Additive Strategies

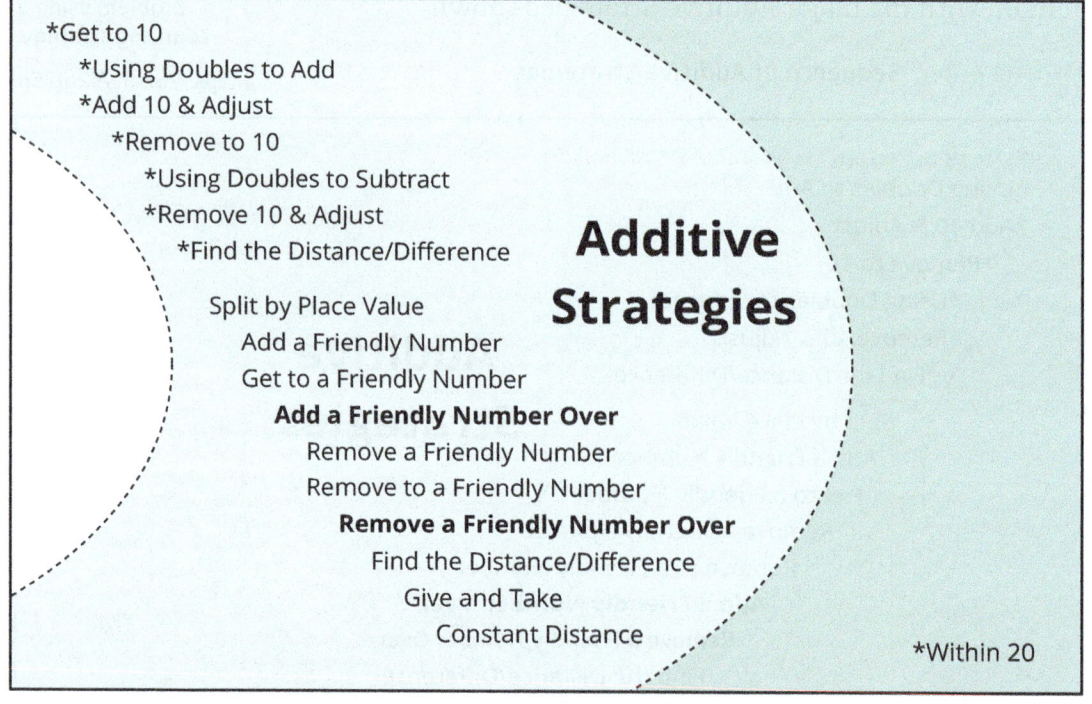

Source: Adapted from Math Is Figure-Out-Able at https://www.mathisfigureoutable.com/ with CC Attribution-NoDerivatives 4.0 International License

The alternative sequence in Figure 4.5 suggests that you help students develop the over addition strategy (Add a Friendly Number Over) before you start developing subtraction strategies. This means that you teach more addition before you start subtraction.

Both sequences work. If you are not sure which to try, we lean toward the more mixed approach . You might feel more comfortable with the more traditional addition-first approach. Please note that either way, we recommend that you hold off developing the addition strategy of Give and Take until after developing all but one subtraction strategy, Constant Difference.

DEVELOPING ADDITION WITHIN 20

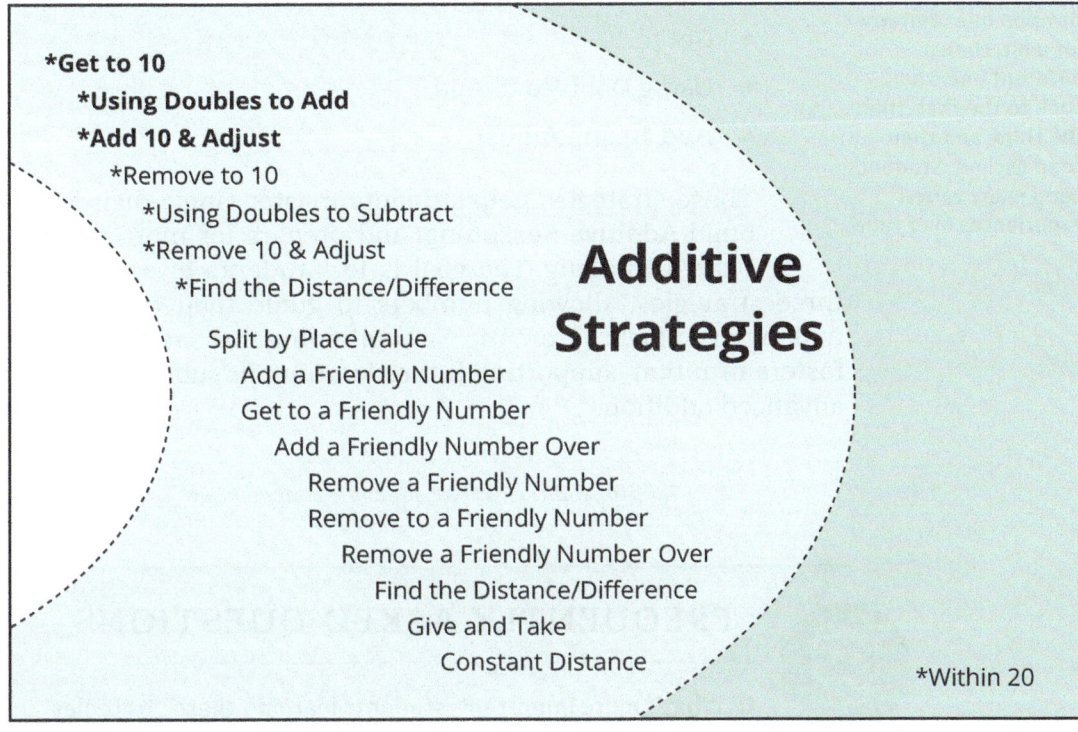

Source: Adapted from Math Is Figure-Out-Able at https://www.mathisfigureoutable.com/ with CC Attribution-NoDerivatives 4.0 International License

When students are solving problems using Counting Strategies, ideally counting on and back, it's time to develop Additive Reasoning by building addition and subtraction strategies for single-digit numbers.

Building Additive Reasoning for addition within 20 is not about having students rote-memorize facts. It's about developing certain important mental connections that make strategies become natural outcomes—intuitive choices based on well-traveled logical mental paths. These neural paths are formed as we give students opportunities to wonder about and notice patterns, try using those patterns, shift and refine that use, make generalizations, experience the teacher making those thoughts/relationships/connections/patterns visible (when my brain does that, it can look like that), refine the words and visual images to fit old and new situations, and repeat.

> **TIP**
> If many of your students are Counting On and Back but some of your students are still Counting All, it is still appropriate for you to begin building Additive Reasoning. Work with students where they are, giving them problems to solve and always nudging them to more sophisticated reasoning with your questioning.

> *It's about developing certain important mental connections that make the strategies become natural outcomes—intuitive choices based on well-traveled logical mental paths.*

> **TIP**
>
> Develop one strategy for a bit, then a different one, then back to the first, then the third, and then keep cycling. Students need many varied experiences over time.

There are three major strategies to actively/purposefully help students develop for single-digit addition:

- Get to 10
- Using Doubles to Add
- Add 10 and Adjust

These strategies help students master single-digit facts, build Additive Reasoning, and prepare for more sophisticated thinking. The goal is for students to develop all three strategies, allowing numbers to guide their choice and making most facts automatic. Strengthening these relationships fosters intuition, supporting future learning in subtraction and advanced addition.

Helpful addition words: addend + addend = sum

FREQUENTLY ASKED QUESTIONS

Q: What's more important, students learning these strategies or students having the facts automatized?

A: This is a false dichotomy. We can have our cake and eat it, too! The best way for students to automatize the single-digit addition facts is for them to actively develop these strategies and use them often. Then, we get not only students with facts at their fingertips but also students who own strategies that grow up into the multi-digit strategies, with all of the mental relationships and connections those bring. Best of both worlds!

FREQUENTLY ASKED QUESTIONS

Q: My students are all over the place. What do I do if some of my students are counting all, some are counting on, and some are using Additive Reasoning?

A: You're not alone. It will always be true that some students will come to you with more or less experience, and some students will need more experiences than others. The good news is that you can help students develop from where they are. Ask questions. Help students make sense of the problems. Encourage them to use what they know. Represent their thinking to make it visible. Help students to notice and use patterns. Celebrate their reasoning. Repeat.

THE GET TO 10 STRATEGY

> **Single-digit Addition**
>
> **Get to 10**
>   Using Doubles to Add
>   Add 10 & Adjust

Source: Adapted from Math Is Figure-Out-Able at https://www.mathisfigureoutable.com/ with CC Attribution-NoDerivatives 4.0 International License

The Get to 10 strategy means starting with one addend, adding part of the other addend to get to 10, and then adding the rest. When students decide to start with one addend and add the other addend in chunks, they end up with 10 plus the leftovers, a teen number.

For example, to add 8 + 7, a student realizes that 8 is close to 10, so add 8 + 2 to get to 10, then think about what's left to add, 7 − 2 = 5. They then add that leftover 5 to 10 and end up with 10 + 5 = 15.

Similarly, a student finding 9 + 5 reasons to get from 9 to 10 by adding 1. They were supposed to add 5, so they need to add 4 more, 10 + 4 = 14.

8 + 7   8 + (2 + 5) = (8 + 2) + 5 = 10 + 5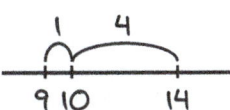

9 + 5   9 + (1 + 4) = (9 + 1) + 4 = 10 + 4

In general, the Get to 10 strategy can look like:

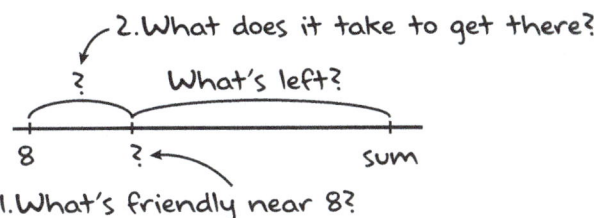

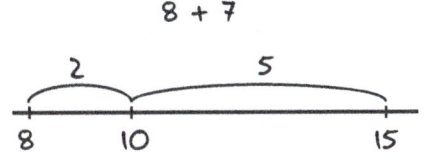

TIP

When you see a student start counting on by ones to add 8 + 7, ask, "Do you know anything you could use? Anything about 8?" If the student pauses too long, ask, "Can you get from 8 to 10?" Pause. "How much from 8 to 10?" Pause. If the student replies with 2, continue with "So if 8 and 2 is 10, how can that help you with 8 and 7?" Pause. "How much more do you need to add? You've added 2." When the student says 5, say, "You have 5 more to add. What is 10 and 5?"

Inherently important in this strategy is the notion of teen numbers, that 10 together with a single-digit number creates a teen. Remember from Chapter 2 that I use the word *teen* to describe the numbers 11–19.

The Get to 10 strategy works well for problems where one addend is close to 10 and the other addend is large enough to make the sum over 10. Examples include 7 + 5, 9 + 6, and 8 + 7.

MODELS TO BUILD THE GET TO 10 STRATEGY

While building the Get to 10 strategy, use both a number rack and an open number line to represent student thinking. Start giving students experience with the discrete number rack model, where students who are still counting one by one are supported, and students who are ready to reason with bigger chunks of numbers are challenged.

8 + 5

8 on the top and 2 on bottom

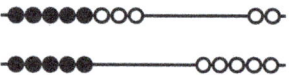

Move 2 to make 10 on top. Compensate on the bottom.

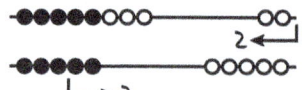

End up with 10 + 3

8 + 5 = 10 + 3 = 13

TIP

As students are developing the Get to 10 strategy, use the structure of the number rack and open number line to highlight the regularity of partners of ten and make up of teen numbers. The first jump always shows the partners of ten, and the next jump shows ten and the rest, which is a teen.

Transition to an open number line, putting tick marks where you need them.

8 + 5

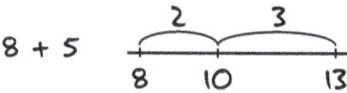

HOW TO TEACH THE GET TO 10 STRATEGY

To help students both get better at and generalize using the Get to 10 strategy, engage students in Problem Strings like the following (Table 4.1).

TABLE 4.1 • Problem String Using the Get to 10 Strategy

PROBLEM	TEACHER
7 + 3	"What is 7 plus 3?" *You can represent this on a number rack or an open number line.*
7 + 5	"Did anyone use the problem before? Could you? How?"
9 + 1	*Repeat.*
9 + 6	"Did anyone use the problem before? How?"
6 + 4	*Repeat. Quickly.*
6 + 5	"Did anyone use 6 + 4 to help? How? It seems helpful to get to that friendly 10 and then add the rest."
8 + 6	"Could you make up your own helper for this one, something friendly? Sure seems helpful to get to that friendly 10 and then add the rest."

Here's a sample final display for this Problem String on a number rack:

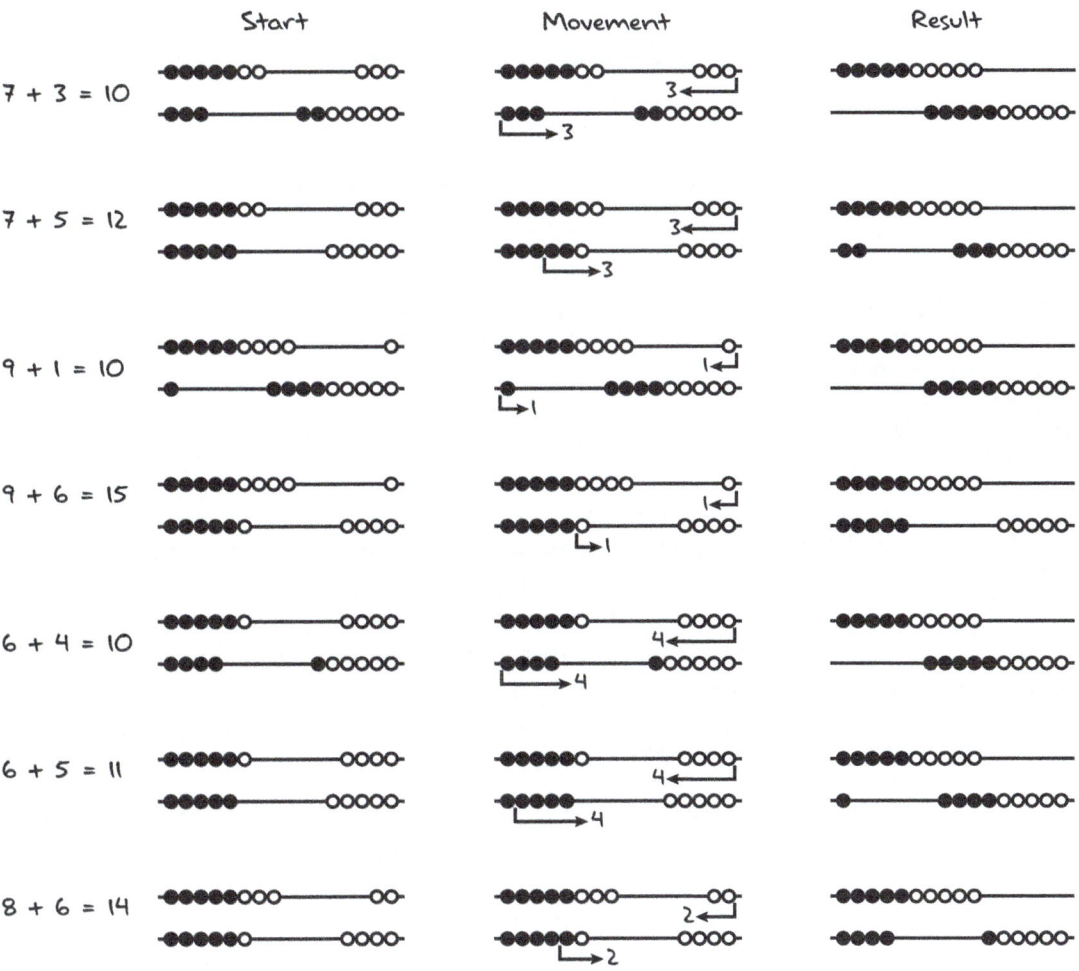

You could do the same kind of Problem String a few days later, this time modeling on an open number line.

Here's a sample final display for this Problem String on a number line:

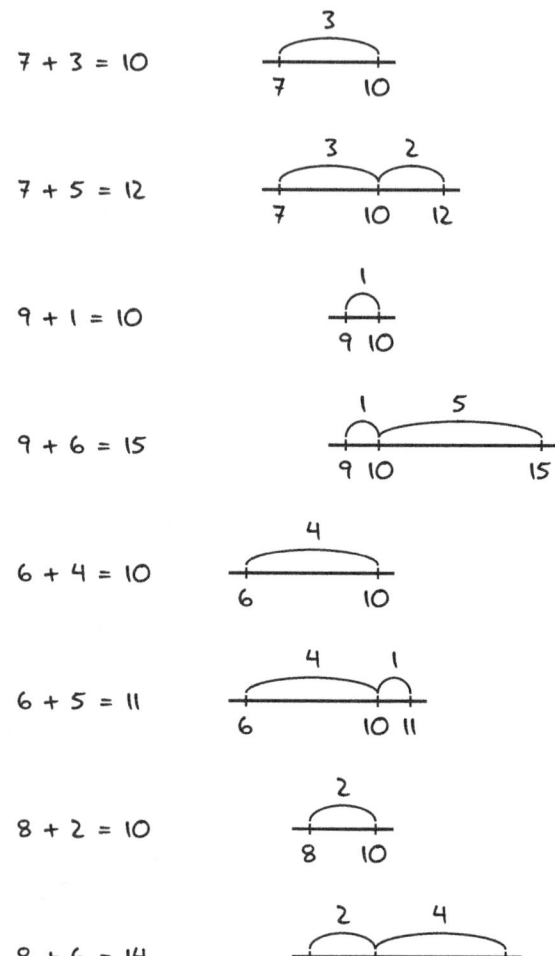

Download a handy PDF for this Problem String.

https://qrs.ly/nrgl25q

TIP

Play I Have, You Need before facilitating Get to 10 Problem Strings. This can help make the partners of 10 top of mind for students and increase the chance that those partners will occur to students during the string. Don't be too pointed about it, not "Remember these partners because you will use them in the Problem String" but rather "Nice work with these helpful partners of 10. I wonder if they will be handy in today's work...." I Have, You Need is discussed further in Chapter 8.

The instructional routine I Have, You Need is superb for getting students ready for and reinforcing the Get to 10 strategy. As students find the partners of 10 repeatedly in this routine, those partners become automatic. Students begin to consider those relationships within other problems, like 8 + 2 within 8 + 5 or 7 + 3 within 7 + 4, and they can use what they know to get to 10 and then add the rest.

Tasks where we purposefully plan the numbers can also help students develop this strategy. Find students to share who combined the partners of 10 first. Discuss how efficient it can be to find those partners of 10 and then just add the rest, a 10 and a single-digit, making a teen.

Try problems like the following:

> The class found 7 markers, 6 pencils, and 4 pens. How many writing tools did they find?
>
> A sports club had donations of 7 baseballs, 5 basketballs, and 3 footballs. How many balls were donated?

> ### TRY IT
>
> Use a Get to 10 strategy for the problem 8 + 6 and represent it with a number rack and an open number line.

TIP

If you notice a student using Counting Strategies who is ready to begin to reason additively, wonder aloud about how the partner of 10 might help. For example, for the problem 9 + 7, ask, "I wonder if thinking about 10 might help? Are we close to ten?"

IMPLICATIONS OF THE GET TO 10 STRATEGY FOR DEVELOPING MATHEMATICAL REASONING

Students will continue to use an extended version of the Get to 10 strategy within multi-digit addition problems. This more sophisticated version is the Get to a Friendly Number Strategy (see Chapter 7), where the friendly numbers are often a multiple of 10.

For example, to add 18 + 5, students can think about getting 18 to the next friendly multiple of 10, which is 20. 18 + (2 + 3) = (18 + 2) + 3 = 20 + 3 = 23.

And it's similar with bigger numbers, like 48 + 26: 48 + (2 + 24) = (48 + 2) + 24 = 50 + 24 = 74.

And with even bigger numbers, students will use partners of 100 to get to the next friendly 100, like 475 + 366 = 475 + (25 + 341) = (475 + 25) + 341 = 500 + 341 = 841.

THE NEXT TWO MAJOR STRATEGIES

The next two major addition strategies to develop in students are the Using Doubles and Add 10 and Adjust strategies. Both of these strategies mean that students are using a fact they know and adjusting.

TIP

React to students' use of strategies with an asset-based perspective. For example, if you are working on the Get to 10 strategy and a student solves 7 + 6 using doubles, refrain from comments like "No, you're supposed to get to 10 today. That's what we're working on." Instead, accept the strategy, keep the conversation about reasoning, and nudge toward the target strategy, "Nice use of doubles! I can understand your thinking. I wonder, do you understand what some of your classmates are doing? They are using 7 + 3 to help. Does that make sense to you?" Students need to feel safe experimenting with their thinking. They will not feel safe if the "wrong" thinking is shot down, especially if they have chosen a different strategy that also works well.

Single-digit Addition

Get to 10
Using Doubles to Add
Add 10 & Adjust

Source: Adapted from Math Is Figure-Out-Able at https://www.mathisfigureoutable.com/ with CC Attribution-NoDerivatives 4.0 International License

Unlike the Get to 10 strategy, the next two strategies require more planning ahead. With Get to 10, students keep one addend whole and begin to add the other addend in pieces that make sense. This requires less thinking ahead because you can add a chunk of one addend before you have fully decomposed it. Both Using Doubles to Add and Add 10 and Adjust strategies require students to consider a helpful relationship and adjust from there.

Because of this, start developing the Get to 10 strategy first, then work on the next two strategies. You can work on both at the same time, but you may find it helpful to work on one for a while, then the other, and continue to alternate as the numbers get a bit more complex, cycling back often to continue to work on the Get to 10 strategy. Then begin to compare the three strategies to help students get better at each.

Let's look at these two important strategies.

THE USING DOUBLES TO ADD STRATEGY

Single-digit Addition

Get to 10
Using Doubles to Add
Add 10 & Adjust

Source: Adapted from Math Is Figure-Out-Able at https://www.mathisfigureoutable.com/ with CC Attribution-NoDerivatives 4.0 International License

The Using Doubles strategy means recognizing that the problem at hand is related to a familiar double, deciding how the problem is related to the double, and adjusting accordingly.

For example, to add 8 + 7, a student realizes that they know 7 + 7 = 14. Since 8 is 1 greater than 7, 8 + 7 is 1 greater than 7 + 7. So 8 + 7 = 7 + 7 + 1 = 14 + 1 = 15.

Similarly, a student finding 8 + 7 could reason that they know 8 + 8 = 16. Since 7 is 1 less than 8, 8 + 7 is 1 less than 8 + 8. So 8 + 7 = 8 + 8 − 1 = 16 − 1 = 15.

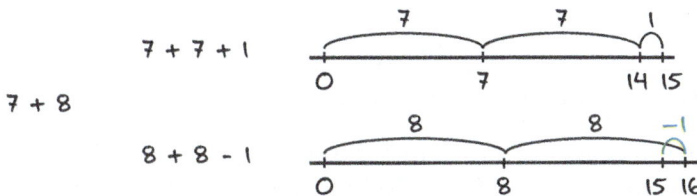

Inherently important in this strategy is the notion that adjusting one of the addends results in adjusting the sum in the same way.

The thinking behind the Using Doubles to Add strategy can look like:

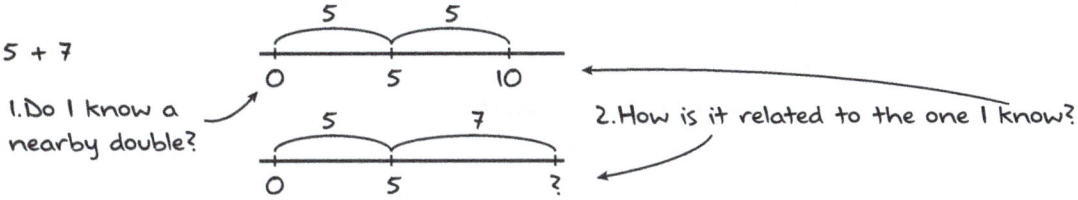

The Using Doubles strategy works well for problems where the addends are close to each other and a nearby double is familiar. For example, if a student knows that 8 + 8 is 16 and recognizes that 9 is one more than 8, they can start with 8 + 8 = 16, adjust up 1 so 8 + 9 is 1 more. 8 + 9 = 8 + 8 + 1 = 16 + 1 = 17.

MODELS TO BUILD THE USING DOUBLES TO ADD STRATEGY

Just like with the Get to 10 strategy, use both a number rack and an open number line.

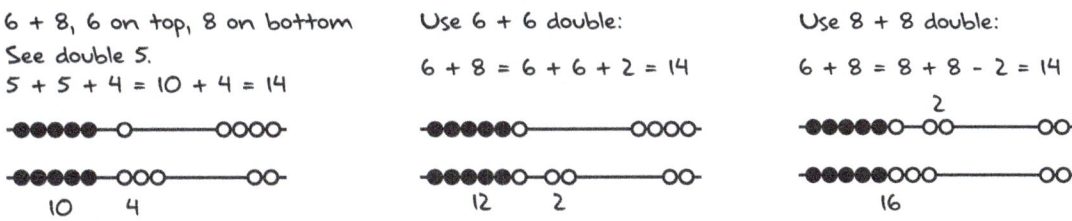

Chapter 4 • The Major Strategies for Addition Within 20

You can show the first addend as a jump from zero to emphasize the doubles visually.

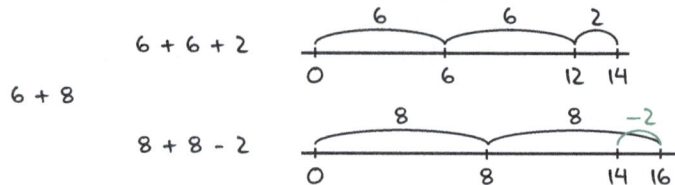

HOW TO TEACH THE USING DOUBLES TO ADD STRATEGY

To help students both generalize and get better at Using Doubles, engage students in Problem Strings like the following (Table 4.2).

TABLE 4.2 • Problem String for the Using the Doubles to Add Strategy

PROBLEM	TEACHER
7 + 7	"What is 7 and 7?" *You can represent this on a number rack or an open number line.*
6 + 7	*Repeat.* "Did anyone use the problem before? Could you? How?"
6 + 6	*Quickly.* "We've been working on doubles, haven't we?"
6 + 8	"Did anyone use any problems from before? How?"
8 + 8	*Repeat.*
9 + 8	"How did you use the problem before?"
8 + 9	"How could you use a double to help?"

Here is a sample final display for this Problem String using a Number Rack:

$7 + 7 = 14 = 5 + 5 + 4 = 10 + 4$

$6 + 7 = 13 = 10 + 3 = \underline{7 + 7 - 1}$

$6 + 6 = 12 = 10 + 2$

$6 + 8 = 14 = \underline{6 + 6 + 2}$

$8 + 8 = 16 = 10 + 6 = 6 + 6 + 4$

$9 + 8 = 17 = 10 + 7 = \underline{8 + 8 + 1}$

$8 + 9 = 17 = 9 + 8 = \underline{9 + 9 - 1}$

Here's a sample final display for this Problem String on an open number line:

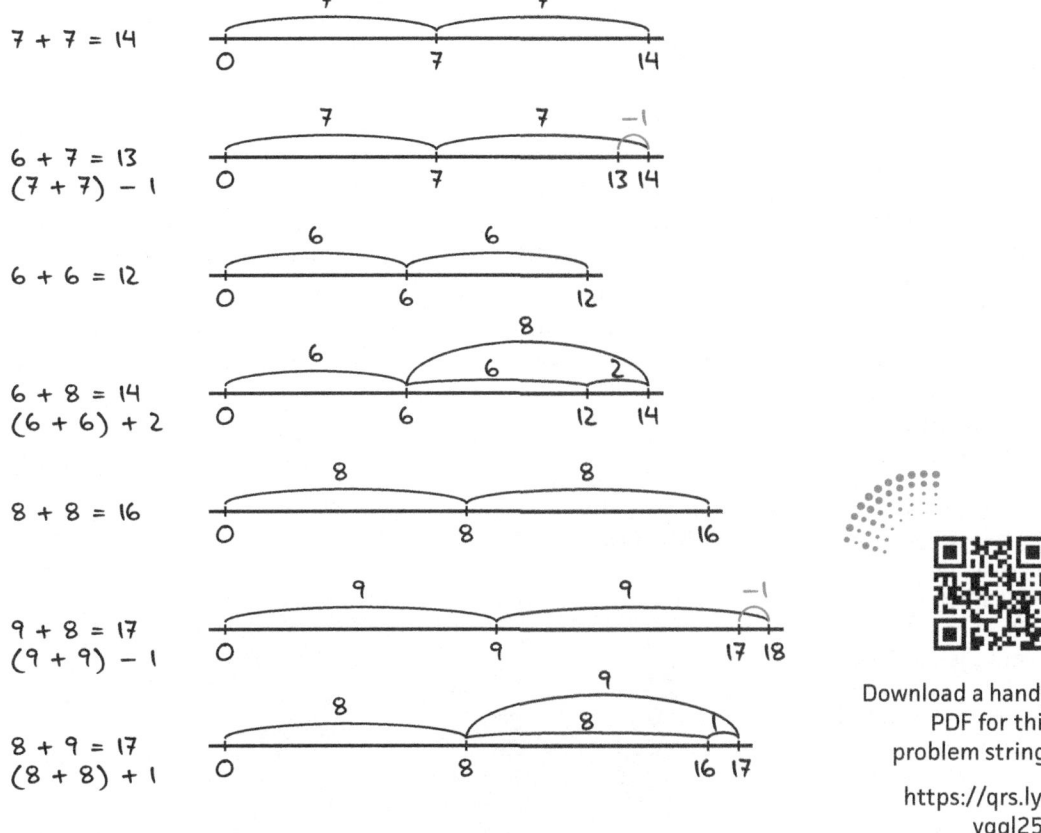

Download a handy PDF for this problem string.

https://qrs.ly/vggl25r

The Problem String at the beginning of this chapter is focusing on building the Using Doubles to Add strategy.

A simple but useful routine to help students develop a repertoire of doubles is to double aloud with students. See Chapter 9 for more about a doubling routine.

Tasks in which we purposefully plan the numbers for problems can also help students develop this strategy.

> Miri had 6 markers and 6 pencils. How many writing tools did she have?
>
> Oops, she just realized she only has 5 pencils. How many writing tools does she actually have?
>
> Daniel found 7 large marbles and 7 small marbles. How many total marbles in his collection?
>
> Wait, he recounted and actually has 7 large and 8 small. How many marbles does he have?

TIP

If students are reasoning about a problem in a way that doesn't match the target strategy but does use Additive Reasoning, for example, using 7 + 6 for 7 + 8 because they love 7 + 6, accept their strategy and ask questions to help nudge toward the target strategy. For example, if you are working on the Using Doubles strategy and a student solves 7 + 6 by getting to 10, refrain from comments like "No, you're supposed to use doubles today. That's what we're working on." Instead, accept their strategy, keep the conversation about reasoning, and nudge toward the target strategy: "Great way to get to 10 and add the leftovers! I can understand your thinking. I wonder, do you understand what some of your classmates are doing? They are using 6 + 6 or 7 + 7 to help. Does that make sense to you?" Remember that it's not just about the answer; it's about building multiple connections.

IMPLICATIONS OF THE USING DOUBLES TO ADD STRATEGY FOR DEVELOPING MATHEMATICAL REASONING

One of the things that makes Using Doubles an interesting strategy is that it can help build toward the sophisticated equivalence strategy of Give and Take in multi-digit addition (Chapter 6). At a young age, students can use doubles with a Give and Take strategy. For instance, they can use 7 + 7 to help with a problem like 6 + 8 by giving 1 from the 8 to the 6 to make 7 + 7. They are thinking ahead, creating an equivalent problem that is easier to solve, and simultaneously considering what happens to the 6 if they take 1 from the 8. This thinking ahead, equivalence, and simultaneity are huge leaps forward in sophistication of thinking. Helping students begin to grapple with them early on will help them Give and Take with doubles with multi-digit problems. We'll discuss more about Giving and Taking in Chapter 6.

THE ADD 10 AND ADJUST STRATEGY

Single-digit Addition

Get to 10
Using Doubles to Add
Add 10 & Adjust

Source: Adapted from Math Is Figure-Out-Able at https://www.mathisfigureoutable.com/ with CC Attribution-NoDerivatives 4.0 International License

The Add 10 and Adjust strategy requires students to think ahead, realizing they can add a bit too much and then adjust back. Students start with one addend whole and decide it would be efficient to add a long jump that is too big because then they can adjust by subtracting off the extra.

For a problem like 8 + 9, think 8 + 10 = 18. But we added 1 too much, so now we need to back up 1 from 18 to 17. So, 8 + 9 is 17.

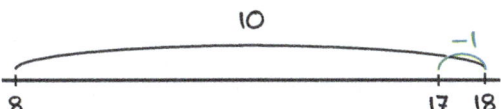

The Add 10 and Adjust strategy uses and helps develop the idea of any single-digit plus 10, the teen numbers. It also has students grappling with what comes before a number. Students have been counting up—this gives them the opportunity to reason about going backward.

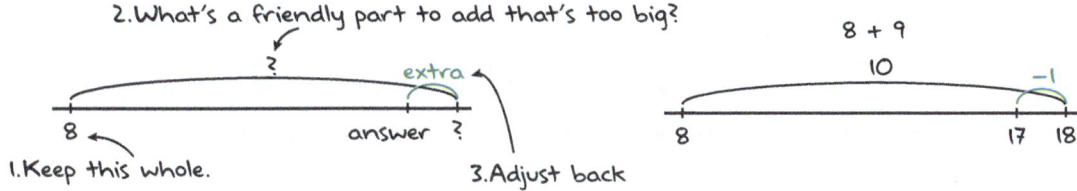

The Add 10 and Adjust strategy is great for adding 9 or 8 to single-digits greater than 2 or 3. For example, 5 + 9 can be thought about as 5 + 10 and then remove 1, so 5 + 10 − 1 = 15 − 1 = 14. Or 7 + 8 can be thought about as 7 + 10, then subtract 2, so 7 + 10 − 2 = 17 − 2 = 15.

This strategy is not that useful for problems where addends are small, like 13 + 4. Adding 13 + 10, then backing up 6 might be fun to play with, but it's not efficient.

MODELS TO BUILD THE ADD 10 AND ADJUST STRATEGY

Just as with the Get to 10 and Using Doubles to Add strategies, use number racks and open number lines to represent student thinking.

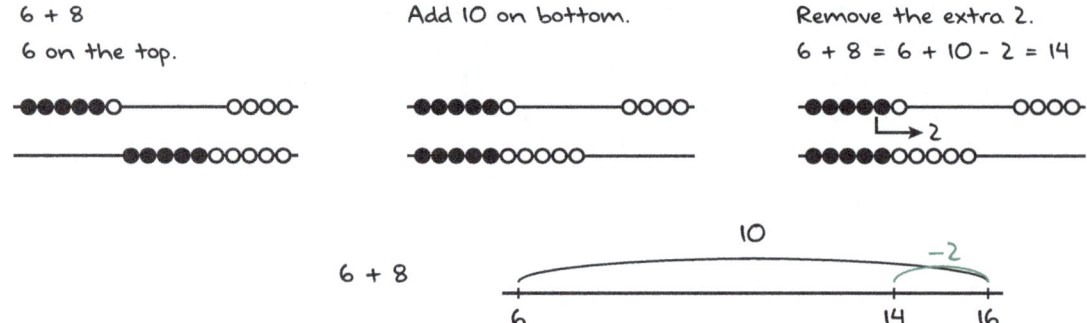

HOW TO TEACH THE ADD 10 AND ADJUST STRATEGY

To help students both generalize and get better at adding the friendly 10 and adjusting back, engage students in Problem Strings like the following (Table 4.3).

TABLE 4.3 ● Problem String Using the Add 10 and Adjust Strategy

PROBLEM	TEACHER
9 + 10	"What is 9 and 10? *You can represent this on a number rack or an open number line.*
9 + 9	*Repeat.* "Did anyone use the problem before? Could you? How? Why back, aren't we adding?"
6 + 10	*Quickly.* "We've been working on adding 10 to any number, haven't we?"
6 + 9	"Did anyone use the problem before? How? Adding a bit too much and then adjusting?"
7 + 10	*Quickly.*
7 + 8	"How did you use the problem before? How is 7 + 10 related to 7 + 8? Why did you go back 2?"
8 + 9	"Could you make up your own helper for this problem? Why 8 + 10? Sure seems like it's helpful to add a bit too much and then adjust."

Here's a sample final display for this Problem String on a number rack:

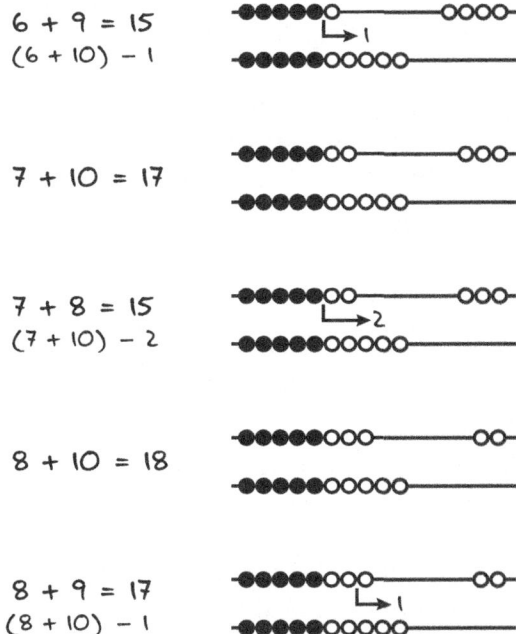

Here's a sample final display for this Problem String on an open number line:

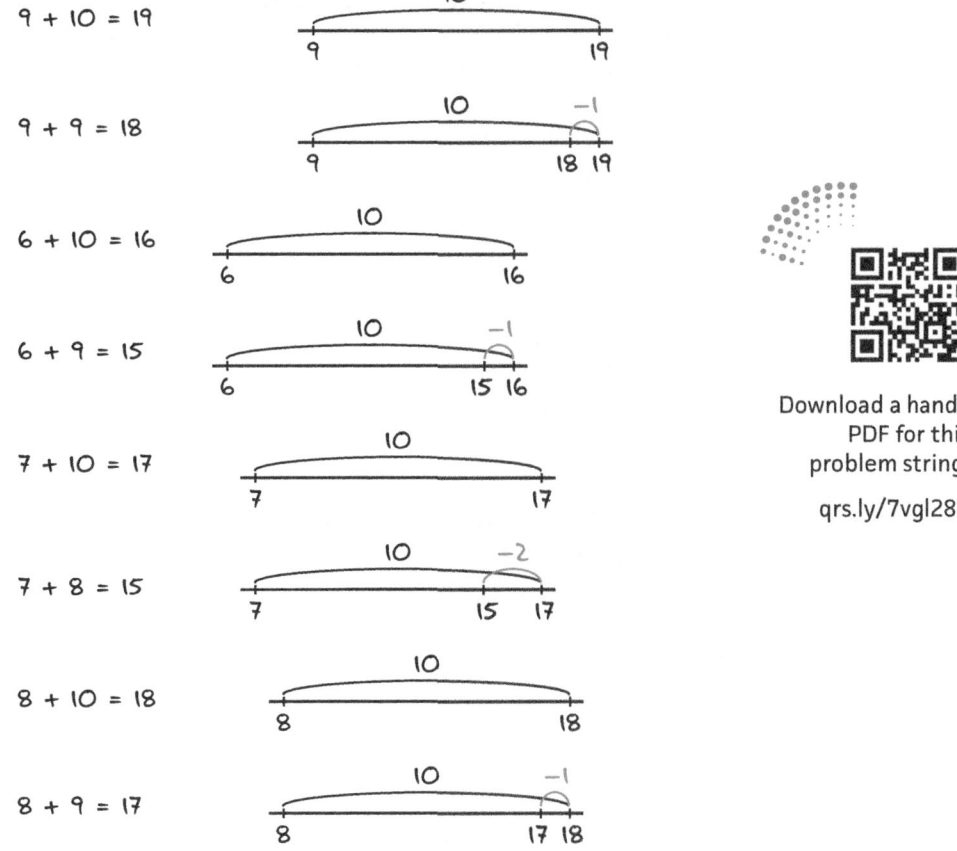

Download a handy PDF for this problem string.

qrs.ly/7vgl284

TIP

When students are learning the teen numbers, say things like "Ah, you mean ten-four and we call that fourteen and write 14."

TIP

When counting back with students, use the word *zero* as the last count rather than "*Blast off.*" This helps students consider the gap, the span, between zero and one as a measurement of 1.

Tasks where we purposefully plan the numbers can also help students develop this strategy.

> Second grade had 5 playground balls, and first grade had 10 playground balls. How many total playground balls did they have for recess?
>
> Actually, they just found out that second grade has 5, but first grade only has 9 playground balls. How many do they actually have?
>
> First grade had 7 skipping ropes. Second grade had 10. How many total skipping ropes?
>
> Wait, second grade recounted and has 9, so first grade has 7, and second grade has 9 skipping ropes. Now how many total skipping ropes?

Because students will be adding single-digit numbers to 10, help students relate that to the meaning of the teen numbers, that a teen is ten and a single-digit. For example, 15, "fifteen," means five-teen, which is like teen-five or ten-five. A better name for fifteen would be ten-five, but we call it fifteen.

Counting backward can be a great way to get students better at adjusting back. Start with a random number, and count back with students. Be sure to start with numbers between multiples, like 12 or 8 rather than always a multiple of 10 or 20. Do this often in short bursts, like when waiting to go out to recess or just back from lunch.

> **TRY IT**
>
> Practice modeling. Move beads on a number rack and draw an open number line to represent a student adding 10 and adjusting back 1 for 7 + 9.

The Add 10 and Adjust is a handy strategy because you keep track as you go and end on the answer.

IMPLICATIONS OF THE ADD 10 AND ADJUST STRATEGY FOR DEVELOPING MATHEMATICAL REASONING

Students will continue to use an extended version of the Add 10 and Adjust strategy within multi-digit addition problems. This later version is the Add a Friendly Number Over Strategy

(Chapter 6), where instead of always adding 10 and adjusting, you add more general friendly numbers, often a multiple of 10, and adjust from there.

For example, to add 28 + 19, students can think about adding the friendly 20. 28 + (20 − 1) = (28 + 20) − 1 = 48 − 1 = 47.

And it's similar with bigger numbers, like 48 + 99: 48 + (100 − 1) = (48 + 100) − 1 = 148 − 1 = 147.

COMPARING THE SINGLE-DIGIT ADDITION STRATEGIES

Each of the three single-digit addition strategies serves different purposes both in building Additive Reasoning and preparing learners for other more sophisticated thinking later (Table 4.4). The Get to 10 strategy helps to establish the primacy of the number 10: if we can get to that 10, we are in a good place. Using Doubles to Add is a good next strategy to explore. Add 10 and Adjust is a bit more sophisticated because it requires some thinking ahead.

TIP

Use strategy language. If you find yourself recognizing that students are using strategies to reason about facts they don't know yet, but you're saying words about speed, you are sending a mixed message. Refrain from comments like "Nice, you did that really fast," calling on the first student who answers or allowing students to call out answers. Instead, try language that is about students' strategies, the way they solved the problem, celebrating the level of sophistication they are using. Try "That seemed really clever to use that double you know." "Once you got to 10, you just made a teen? Very efficient!" "Since you know plus 10, you can just adjust back 1 to add 9? Very handy."

TABLE 4.4 • Comparing Single-Digit Addition Strategies

STRATEGY	DEALING WITH THE ADDENDS	THINKING ABOUT	WHAT IT BUILDS
Get to 10	Keep one addend whole, break the other into the first addend's partner of 10 and the leftovers	*Would getting to 10 help me?*	Partners of 10, decomposing the other addend, Adding 10 and a single-digit number (the meaning of the teen numbers)
Using Doubles to Add	Keep one addend whole, break the other addend into the first addend and the leftovers	*Can I adjust the double I know?*	Doubles, nearness, order of numbers
Add 10 and Adjust	Keep on addend whole and add 10 to make a teen, adjust back the overage	Requires planning ahead, *Can I add a bit too much and adjust back?*	Adding 10 to a single-digit number (meaning of teen numbers), partners of 10 (as they decide how much to adjust back), and counting backward

> **TRY IT**
>
> Find 6 + 8 using each of the three strategies. Represent the strategy with at least one model.

All of these three strategies are excellent for students to automatize the single-digit addition facts and are the needed preparation to develop subtraction strategies and the major multi-digit addition strategies.

> **FREQUENTLY ASKED QUESTIONS**
>
> **Q:** What do I do if students know the strategies but are not using them?
>
> **A:** If you have done tasks and Problem Strings and your students can discuss the strategies and how they make sense but you still find them counting to find the sum of single digits, you can do a few things:
>
> 1. Concentrate on keeping the conversations about strategy—how did you find that? If you counted, how could you have used relationships?
> 2. Gently but firmly tell students that their job is to use reasoning, not to count. Celebrate when they use a strategy. Positively reinforce.
> 3. Make it about sophistication, not speed. Instead of, "Wow, you did that really fast," try, "Nice thinking, that seems really clever."
> 4. Think aloud as you use these strategies, modeling for students how to keep them front of mind.

"You did what?" I asked Kim. We had been working with a school in Austin for a few years, coaching teachers. On this day, Kim had gone to work with first-grade teachers.

"I told them to stop that nonsense," she stated and grinned. "The kids just needed to be clear what their job is now."

The day before, an excellent first-grade teacher had told Kim, "We've done so much work with our students, the kids know the strategies, but some of the kids are not using them. They're counting fast to add and subtract; they're not using relationships."

So, that morning, Kim had gathered this small group of students on the floor in the corner of the room and quickly did an interview to check where kids were on the landscape of learning (Fosnot & Dolk, 2001), making sure they actually did own the strategies:

> After I was satisfied that these students did indeed own the strategies, I gave them a problem. 'We've got 6 students here. If 8 students join us, how many students will we have?' The students instantly, confidently, put up fingers and began counting by ones. I gently reached over and sort of put my hand on their fingers. 'Stop counting. You know several strategies. This is when you *use* them.' They looked at me, paused, and started using strategies. We briefly discussed whether they liked the Using Doubles or Get to 10 strategy they had used, and then I gave them another problem. 'Remember, this is when you use those strategies.' Little light bulbs went off. They connected that those relationships they had been building with games and Problem Strings could be and should be used when they were actually solving problems. Many of their classmates had already realized it, but for these few students, they needed explicit help making the connection. You've got to find out who knows things, and tell them to stop doing the nonsense. Politely!

Kim wasn't shaming or asking students to mimic something they were supposed to have memorized. She positively encouraged the students to reason mathematically: "Stop counting. We already know you can do that. It's time to use the strategies you've been learning. You can do it."

Kim's actions were a gentle reminder—use what you know, the strategies you own. When you as a teacher know your content and your kids, you know when to respectfully nudge students toward more sophisticated thinking. I am not suggesting that you tape students' fingers together. This happened to our colleague Kourtney's sister, whose name ironically is Kym. But Kym didn't know what else she could do, just that using her fingers was now impossible because they were taped together. The problem is not counting on fingers: it's counting one by one when a student is ready for more.

The biggest take away here is that we must connect the strategies that we are building in Problem Strings to the other parts of math class and to problems in real life if we want students to use them.

By the end of first grade, we want students to have created a dense set of relationships and multiple ways to reason about facts. We want them to have enough experiences reasoning with relationships that even if a particular fact isn't automatic, they can refigure it quickly. However, speed isn't the goal in "knowing your facts." Automaticity is about creating well-traveled mental pathways that a student can follow without veering off into dead ends. It's about compressing multiple relationships that are well understood so that students can have the facts at their fingertips but also uncompress and dive into the underlying relationships when needed. Whether a student "just knows" 8 + 5 or refigures it with minimal effort, the mental energy required to solve it is minimal, and the student's mental capacity is available to work on building other connections.

Conclusion

Learning the single-digit addition facts is important. The best way to help student automatize the facts and learn important additive relationships is by developing the major strategies. These facts, relationships, and strategies create the best foundation for success with subtraction and multi-digit operations. Our goal is to encourage students to continue making new mental connections, using those relationships to solve problems so that students build more sophisticated reasoning. Addition is figure-out-able!

Discussion Questions

1. Which of the major single-digit addition strategies do you find yourself gravitating toward (tend to think of first)? Which is less obvious to you?

2. Which of the single-digit strategies involve using partners of 10? How?

3. Enact with a thought partner: You're working with a student who does not just know the answer to 7 + 8. Choose from the following prompts and create a plausible back-and-forth focusing pattern of questioning: "Do you want to think about 7 + 8 or 8 + 7? Do you know something close to 7 + 8? What is friendly that is close to 7? Close to 8? Do you know 7 + 10?"

4. Choose another often missed fact, like 8 + 6. Write your own prompts, and create a plausible back-and-forth focusing pattern of questioning.

5. Interview your students to find which facts they know and which they do not know yet. Make a comprehensive list of the facts on most students' lists. Make a plan to work on those by giving students varied experiences with word problems, Problem Strings, games, and other routines.

CHAPTER 5

The Major Strategies for Subtraction Within 20

FIGURE 5.1 ● The Second Level of Sophistication in Mathematical Reasoning

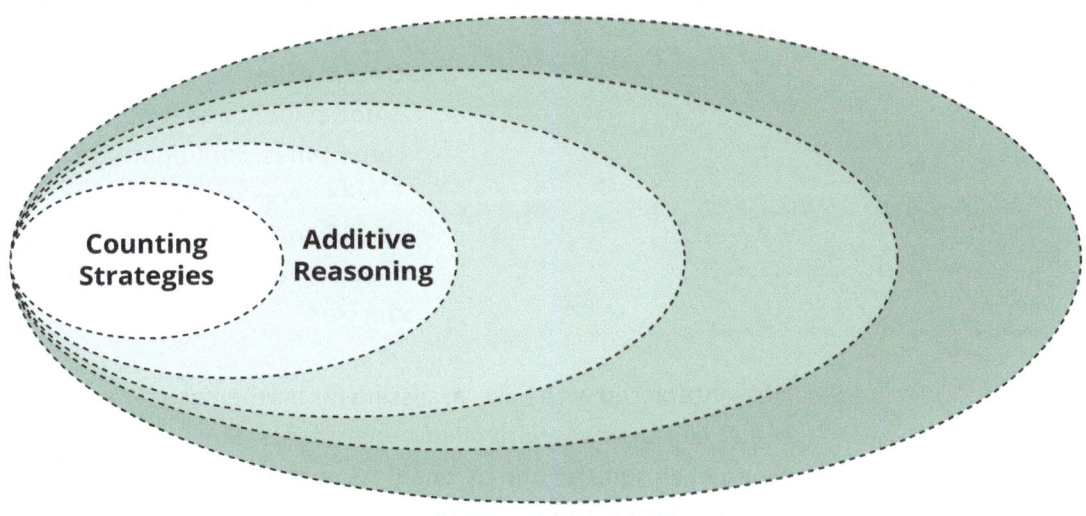

Source: Adapted from Math Is Figure-Out-Able at https://www.mathisfigureoutable.com/ with CC Attribution-NoDerivatives 4.0 International License

Sarah Hempel, a first-grade teacher, circulates around the room as students answer a couple of questions. They had been solving several combining-result unknown problems and were just starting to solve separating result-unknown problems (see Chapter 3).

One of the problems at hand was about squirrels. Sarah noticed that Luke had painstakingly drawn each squirrel from the problem, in the action described (Figure 5.2).

FIGURE 5.2 • Luke's Solution to the Squirrel Problem

What do you think about Luke's strategy? You can see the total 12 squirrels, with the 8 squirrels on the ground and the 4 squirrels running up the tree. This must be direct modeling of the problem, where Luke is acting out the action and then Counting Three Times, right?

But Sarah notices something.

Luke has already drawn the result. In order to draw this picture, he must have already found that there were 8 squirrels left on the ground if 4 had run up the tree. In other words, this picture is not evidence that he counted three times. For Counting Three Times, we would expect to see 12 squirrels on the ground, then 4 crossed out and put on the tree.

Sarah has interacted with Luke, assessing his mathematical reasoning often, just like she does with all of her students. She sees that Luke is drawing each squirrel one by one (less sophisticated) but has the intuition that he has already figured out the answer in an additive way (more sophisticated). So Sarah asks Luke to tell her about his thinking. Luke replies, "Well, yesterday we worked with 8 + 4 = 12. So then 12 minus 8 is 4." As Sarah understands his thinking, she models it using words and equations and tells Luke, "Ah, nice reasoning. When you think that way, you can represent it like this." In Figure 5.3 you can see Sarah's modeling of Luke's strategy with words and equations.

FIGURE 5.3 • Sarah's Modeling of Luke's Actual Strategy

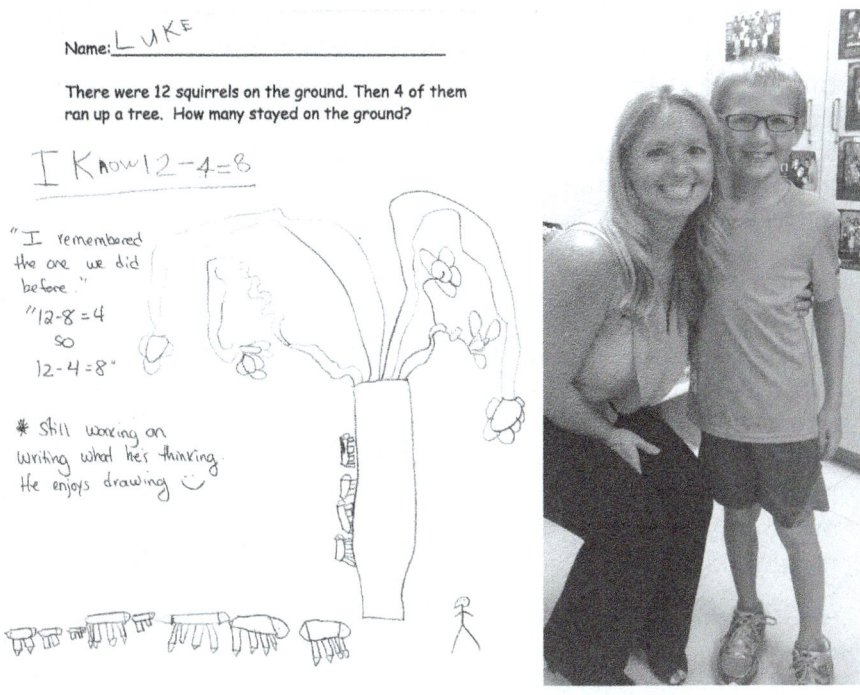

Source: Kim Montague

DEVELOPING SUBTRACTION WITHIN 20

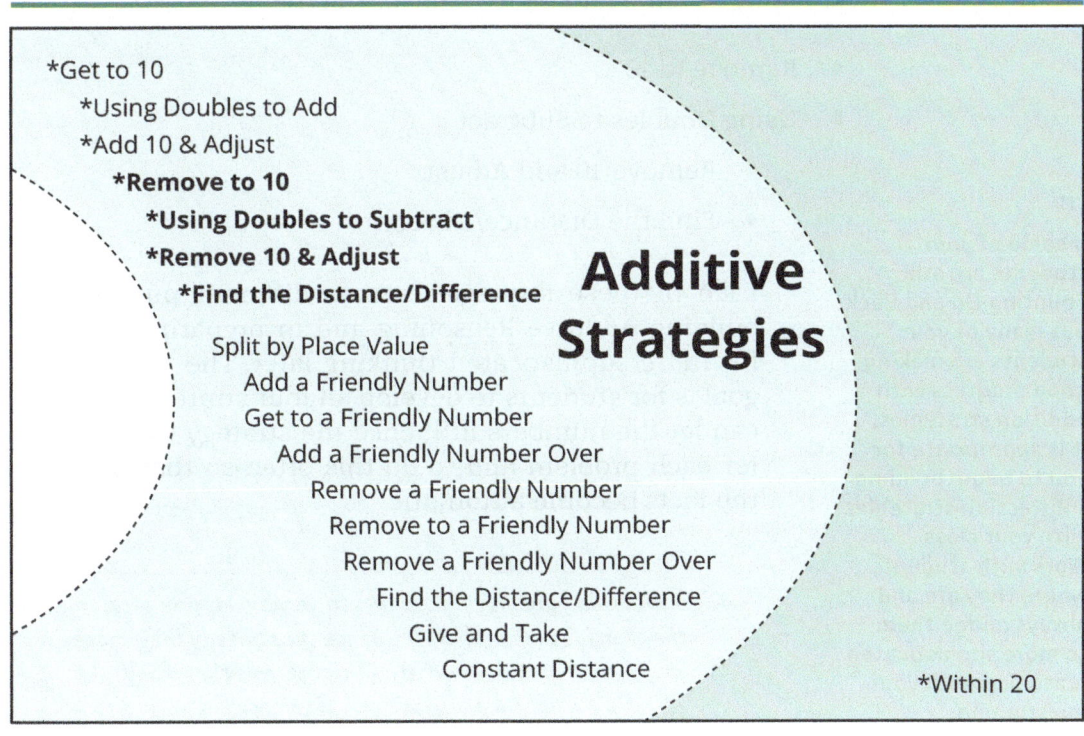

Source: Adapted from Math Is Figure-Out-Able at https://www.mathisfigureoutable.com/ with CC Attribution-NoDerivatives 4.0 International License

Video of a student solving a subtraction problem using a counting strategy.

https://qrs.ly/fegl25z

Video of a student solving a subtraction problem using an additive strategy.

https://qrs.ly/2fgl260

TIP

If some of your students are still Counting On and Back but many of your students are making good progress with addition strategies, it is appropriate for you to begin building subtraction strategies with your class. Work with students where they are and always nudge them to more sophisticated reasoning with your questioning.

When students have begun to develop Additive Reasoning by building the addition strategies, it's time to continue building Additive Reasoning with subtraction within 20. Students might revert to Counting Strategies when you start subtraction. It's often true that you revert to something you know when you encounter something new. Keep developing Additive Reasoning, and they will begin using additive relationships when subtracting.

Building Additive Reasoning for subtraction within 20 is about building on the addition strategies and developing a few major mathematical relationships that lead to students naturally and intuitively using a few important subtraction strategies. Just like with addition, refrain from having students memorize and mimic strategies. Help them continue to develop important mental connections that make the strategies become natural outcomes—intuitive choices based on well-traveled logical mental paths.

Help students continue to develop certain important mental connections that make the strategies become natural outcomes—intuitive choices based on well-traveled logical mental paths.

There are four major strategies to actively/purposefully help students develop for single-digit subtraction:

- Remove to 10
- Using Doubles to Subtract
  - Remove 10 and Adjust
  - Find the Distance/Difference

Each of these strategies serves different purposes in building Additive Reasoning and in preparing learners for more sophisticated thinking later. The overarching goal is for students to develop all four strategies so they can let the numbers influence the strategy they choose for each problem and to do this often so that many of the facts become automatic.

The overarching goal is for students to develop all four strategies so they can let the numbers influence the strategy they choose for each problem and to do this often so that many of the facts become automatic.

Start by developing the Remove to 10 strategy. Then work on one of the next two. Develop one strategy for a bit, then a different one, then back to the first, then the third, and then keep cycling. Students need many varied experiences over time. You will develop Find the Distance/Difference last.

Melisa Williams, a first-grade teacher, uses Problem Strings to develop the Remove to 10 strategy with her first-grade class.

TIP

Remember that the technical subtraction words are *minuend − subtrahend = difference*, I also use *first number* to refer to the minuend and *second number* to refer to the subtrahend.

By the end of the Problem String, students are solving 13 − 5 by thinking about 13 − 3 to get to 10 and then removing the extra 2 to land on 8.

THE REMOVE TO 10 STRATEGY

Melisa and her students reasoning in action.

https://qrs.ly/tpgl261

Single-digit Subtraction
Remove to 10
Using Doubles to Subtract
Remove 10 & Adjust
Find the Distance/Difference

The Remove to 10 strategy entails starting with the first number, subtracting part of the second number to get to 10, then subtracting the rest. As students subtract the second number in chunks, they end up with 10 minus the leftovers, a single-digit number. This last relationship is our good friend, partners of 10.

For example, to subtract 17 – 8, a student realizes that subtracting 8 will bridge 10, so remove 7 first to get to 10, then think about what's left to remove, 8 – 7 = 1. Subtract that leftover 1 from 10 and end up with 10 – 1 = 9.

Similarly, a student finding 15 – 7 reasons to get from 15 to 10 by removing 5. They were supposed to subtract 7, so they need to remove 2 more, so 10 – 2 = 8.

17 – 8   (17 – 7) – 1 = 10 – 1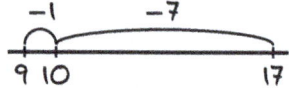

15 – 7   (15 – 5) – 2 = 10 – 2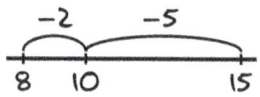

Inherently important in this strategy is the notion of teen numbers, that ten together with a single-digit number creates a teen.

In general, the Remove to 10 Strategy can look like this:

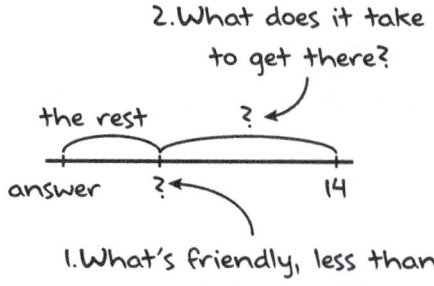
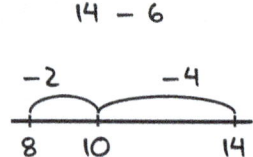

108 Part III • Developing Additive Reasoning

This strategy is based on breaking apart and reassociating the subtrahend in a manner similar to the Get to 10 strategy of addition.

The Remove to 10 strategy works well for problems where the minuend is a teen number and the subtrahend is larger than the ones digit of the minuend. For example, 15 – 8, 15 is a teen number and subtracting 8 requires passing 10. Start at the 15, subtract the 5 to get to 10, then remove the leftover 3, 10 – 3 = 7.

MODELS TO BUILD THE REMOVE TO 10 STRATEGY

Use both a number rack and an open number line to represent student thinking. Start giving students experience with the discrete number rack model, where students who are still counting one by one are supported, and students who are ready to reason with bigger chunks of numbers are challenged.

13 – 5
10 on top, 3 on bottom.

Move 3 on bottom to get to 10.
Move 2 on top to finish the removal.

End up with 8.
13 – 5 = 13 – 3 – 2 = 8

Transition to an open number line, putting tick marks where needed to represent thinking.

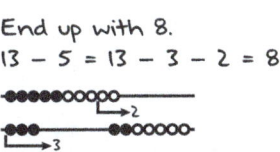

> ### TRY IT
> Solve the problem 16 – 9 using the Remove to 10 strategy. Represent it on a model.

TIP

When working with students who are still counting by ones to subtract 13 – 5, use the models to help nudge removing in bigger chunks. On the number rack, ask, "What could we remove to get to 10? We need to remove 5 total, but what can we see right away that we could start with to get to 10?" On a number line, start at the 13, and ask, "What is a number we could jump to first? What do you know about 13 that could help us here?"

HOW TO TEACH THE REMOVE TO 10 STRATEGY

To help students both get better at and generalize using the Remove to 10 strategy, engage students in Problem Strings like those in Table 5.1.

TABLE 5.1 • Problem String Using the Remove to 10 Strategy

PROBLEM	TEACHER
14 − 4	"What is 14 minus 4? How do you know?"
14 − 6	"What is 14 − 6? Did anyone use the previous problem to help? Could you? How?"
17 − 7	*Repeat.* "How are you thinking about 17 − 7? Teens are a 10 and some ones, aren't they?"
17 − 8	"Did anyone use the 17 − 7 to help us with this one? How?"
12 − 2	*Repeat. Quickly.* "There's that helpful 10 again!"
12 − 5	"Did anybody use the previous problem to help? How?"
16 − 7	"Could you create your own helper problem for this one? It seems helpful to remove to ten and then remove what's left."

Here's a sample final display for this Problem String on a number rack:

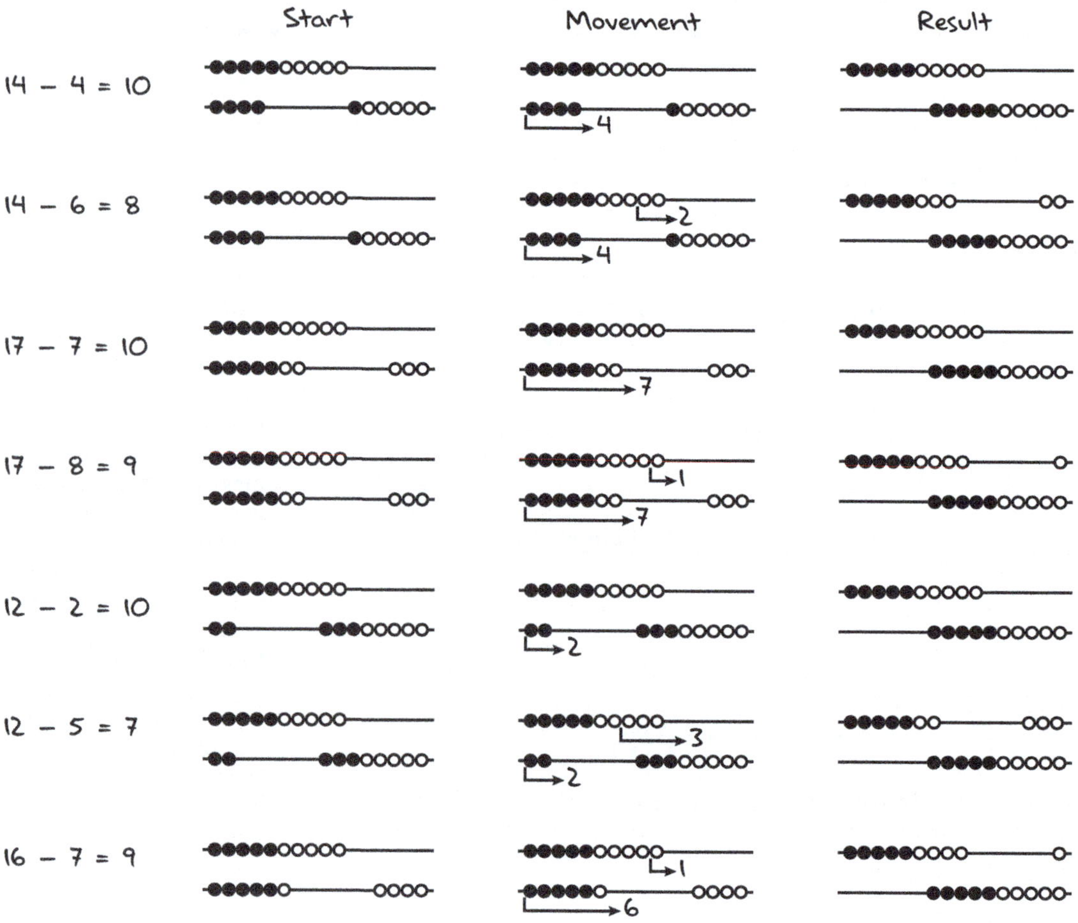

14 − 4 = 10

14 − 6 = 8

17 − 7 = 10

17 − 8 = 9

12 − 2 = 10

12 − 5 = 7

16 − 7 = 9

You could do a similar string a few days later, this time modeling on an open number line. Here's a sample final display for this Problem String on a number line:

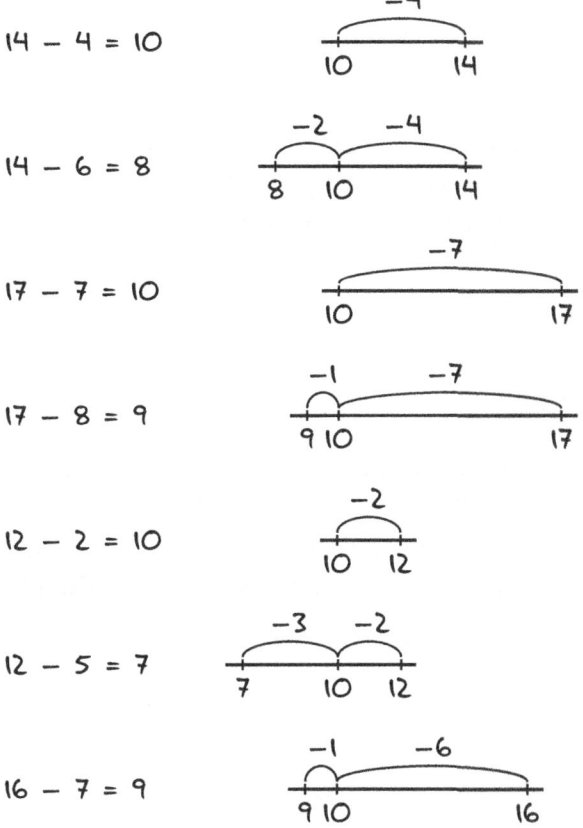

Download a handy PDF for this Problem String.

https://qrs.ly/9igl263

The reasoning for this strategy is very similar to the Get to 10 strategy for addition but in reverse: students first apply their understanding of teen numbers and *then* use partners of 10 to remove the rest. The Teens Remove Single Digits strategy is a precursor to the Remove to 10 strategy because it helps strengthen understanding of teen numbers as a 10 and some 1s. The Remove to 10 strategy starts with removing to 10, or removing the value of the ones digit in the teen number first. If students own their partners of 10, removing the leftovers from 10 becomes a question of applying those partners of 10 to see the answer to the subtraction problem.

Tasks where we purposefully plan the numbers can also help students develop this strategy. Find students to share who removed to 10 first. Discuss how efficient it can be to remove to 10 first, then use partners of 10 to finish the problem.

TIP

Continue to play I Have, You Need as you build subtraction strategies. When the partners of 10 are top of mind, removing from 10 becomes much easier. When working a Remove to 10 Problem String with students, nudge them to think about partners of 10 for that final step: "And what is 10 – 6? Could we think of partners of 10 to help? Would it help if I said 'I have 6, you need . . .'"

FREQUENTLY ASKED QUESTIONS

Question: When students are sharing, what if none of my students removed to 10?

Answer: A favorite teacher move of mine is to then mention that this is how students from a previous year solved the problem and ask the students why they think my student from before might have done it this way. Don't make it sound like they should have known to do that. Just be super curious about why your previous student did. Then maybe ask if anyone wants to try that.

TIP

If you notice a student using Counting Strategies who is ready to begin to reason additively, wonder aloud about if there is an amount they might remove all at once to get to a friendly number. For example, for the problem 14 − 5, ask, "When you think about 14, is there a number you could remove first that is less than 5 and gets you to a friendly number?"

Problems like the following can help build this strategy:

The class had 16 markers, and then 7 of them ran out of ink. How many markers were left?

There were 14 birds sitting on a tree. If 6 of the birds flew away, how many were left on the tree?

TRY IT

Solve 14 − 6 using a Remove to 10 strategy. Model on a number rack and an open number line.

IMPLICATIONS OF THE REMOVE TO 10 STRATEGY FOR DEVELOPING MATHEMATICAL REASONING

Students will continue to use an extended version of the Remove to 10 strategy within multi-digit addition problems. This more sophisticated version is the Remove to a Friendly Number Strategy (see Chapter 7), where the friendly numbers are often a multiple of 10.

For example, to subtract 25 − 7, students can think about removing to the next friendly multiple of 10, which is 20 here: 25 − 7 = (25 − 5) − 2 = 20 − 2 = 18.

And with even bigger numbers, students will use partners of 100 to remove to the next friendly 100, like 425 − 236 = (425 − 25) − 211 = 400 − 211 = 189.

THE NEXT TWO MAJOR STRATEGIES

The next two major subtraction strategies to develop in students are the Using Doubles to Subtract and Remove 10 and Adjust strategies. Both of these strategies mean that students are using a fact they know and adjusting.

> **TIP**
>
> Connect subtraction strategies to their parallel addition strategies to help students use what they have been constructing for addition. For example, "When we were adding, sometimes it was helpful to get to that friendly 10 first and add the rest. Could that idea be useful here with subtraction? Could we remove some chunk to get to 10 first, then remove the rest?"

Single-digit Subtraction
Remove to 10
Using Doubles to Subtract
Remove 10 & Adjust
Find the Distance/Difference

Source: Adapted from Math Is Figure-Out-Able at https://www.mathisfigureoutable.com/ with CC Attribution-NoDerivatives 4.0 International License

Unlike the Remove to 10 strategy, the next two strategies require more planning ahead. With Remove to 10, students keep the minuend whole and begin to subtract the subtrahend in pieces that make sense. This requires less thinking ahead because you can remove a chunk of the subtrahend before you have fully decomposed it. Both Using Doubles to Subtract and Remove 10 and Adjust strategies require students to consider a helpful relationship and adjust from there.

Because of this, start developing the Remove to 10 strategy first, then work on the next two strategies. You can work on both at the same time, but you may find it helpful to work on one for a while, then the other, and continue to alternate as the numbers get a bit more complex, cycling back often to continue to work on the Remove to 10 strategy. Then begin to compare the three strategies to help students get better at each.

> **FREQUENTLY ASKED QUESTIONS**
>
> **Q:** Are you saying it's okay to start with either Using Doubles to Subtract or Remove 10 and Adjust? It doesn't matter which order?
>
> **A:** Correct! Maybe try it one way this year and a different next. Either way, help students develop both strategies over time with many experiences.

Let's look at these two important strategies.

THE USING DOUBLES TO SUBTRACT STRATEGY

Single-digit Subtraction
Remove to 10
Using Doubles to Subtract
Remove 10 & Adjust
Find the Distance/Difference

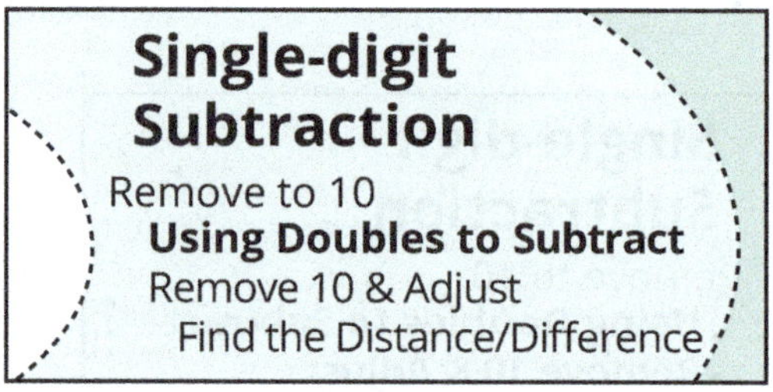

Source: Adapted from Math Is Figure-Out-Able at https://www.mathisfigureoutable.com/ with CC Attribution-NoDerivatives 4.0 International License

The Using Doubles to Subtract strategy means recognizing that the problem at hand is related to a familiar double, deciding how the problem is related to the double, then adjusting accordingly. The strategy works when the minuend is the double, or near the double, of the subtrahend.

For example, to subtract $16 - 7$, a student realizes they know $8 + 8 = 16$ and, therefore, that $16 - 8 = 8$. Since removing 8 would be taking away 1 too many, they need to add 1 back to $16 - 8$. $16 - 7 = 16 - 8 + 1 = 8 + 1 = 9$.

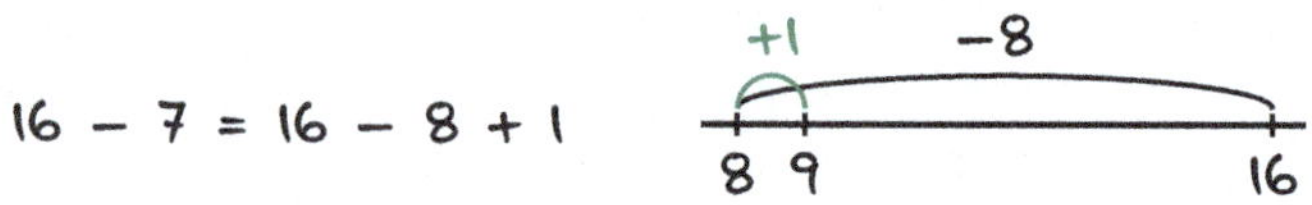

$16 - 7 = 16 - 8 + 1$

Another way to use doubles for 16 – 7 is to think about the double of 7, 14. Then adjust from there up 2.

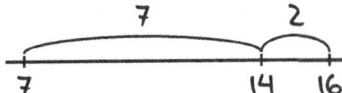

When a solver starts to think about using doubles for a problem, they can consider the different ways that doubles could be used, in one version thinking about half and the other version about double. For example, with 14 – 8, students can consider that since half of 14 is 7, they can use 14 – 7 = 7 then subtract 1 more to get 6. Alternatively, they could think about how 8 + 8 = 16, so since 14 is 2 less than 16, 14 – 8 will be 2 less than 8, or 6. Considering using doubles as both doubles and halves helps the solver build flexibility in problem solving and in doubling and halving.

Inherently important in this strategy is the notion that adjusting the subtrahend up or down from half of the minuend results in adjusting the difference in the opposite way.

The thinking behind the Using Doubles to Subtract Strategy can look like the following:

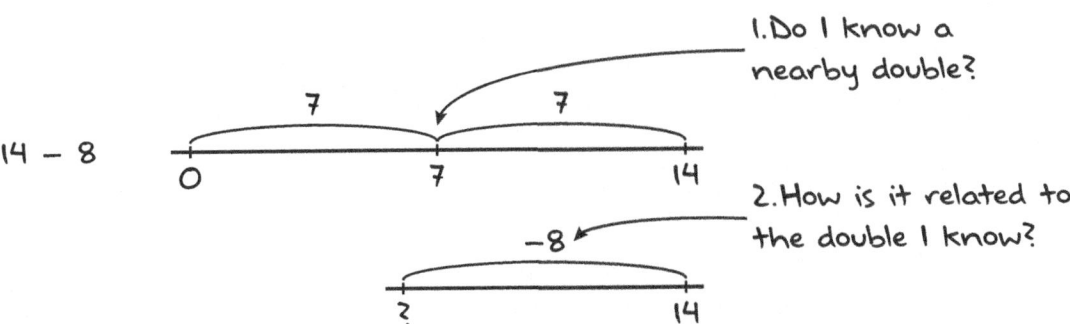

The Using Doubles strategy works well for problems where the minuend is an even number known to the solver as a double and the subtrahend is close to half of the minuend. For example, if a student knows that 8 + 8 = 16 and therefore that 16 – 8 = 8, they could use doubles to solve the problem 16 – 7 in the following way: 16 – 7 = 16 – 8 + 1 = 8 + 1 = 9.

> **TRY IT**
>
> Name three problems that would work well for the Using Doubles to Subtract strategy.

MODELS TO BUILD THE USING DOUBLES TO SUBTRACT STRATEGY

Just like with the Remove to 10 Strategy, use both a number rack and an open number line.

16 − 7

8 on top, 8 on bottom
Remove 7 on bottom
See answer as 8 + 1 = 9

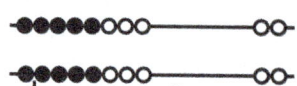

16 − 9

8 on top, 8 on bottom
Remove bottom 8, then
1 more from top row
See answer as 8 − 1 = 7

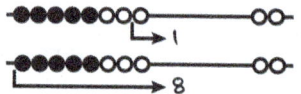

Transition to an open number line, putting tick marks where needed to emphasize relationships.

16 − 7

HOW TO TEACH THE USING DOUBLES TO SUBTRACT STRATEGY

To help students both generalize and get better at Using Doubles, engage students in Problem Strings like the one in in Table 5.2.

TABLE 5.2 • Problem String for the Using Doubles to Subtract Strategy

PROBLEM	TEACHER
8 − 4	"What is 8 − 4? Is that a double you know? Doubles are helpful, aren't they!"
8 − 3	"Can you use 8 − 4 to help you with 8 − 3? How? So did you subtract more this time or less?" "How did that change the answer?"

116 Part III • Developing Additive Reasoning

PROBLEM	TEACHER
8 − 5	"I wonder if there is a problem up here that could help us? Did you subtract too much or too little? How can you adjust?"
14 − 7	"What is 14 − 7?"
14 − 8	"Did anybody use the previous problem to help? Could you?"
14 − 6	"Does knowing 14 − 7 = 7 help us with this one? How?"
12 − 7	"What problem could you use to help you solve 12 − 7? How could you adjust? It seems helpful to think about the doubles we know to help us solve problems that are near doubles."

Here's a sample final display for this Problem String using a Number Rack:

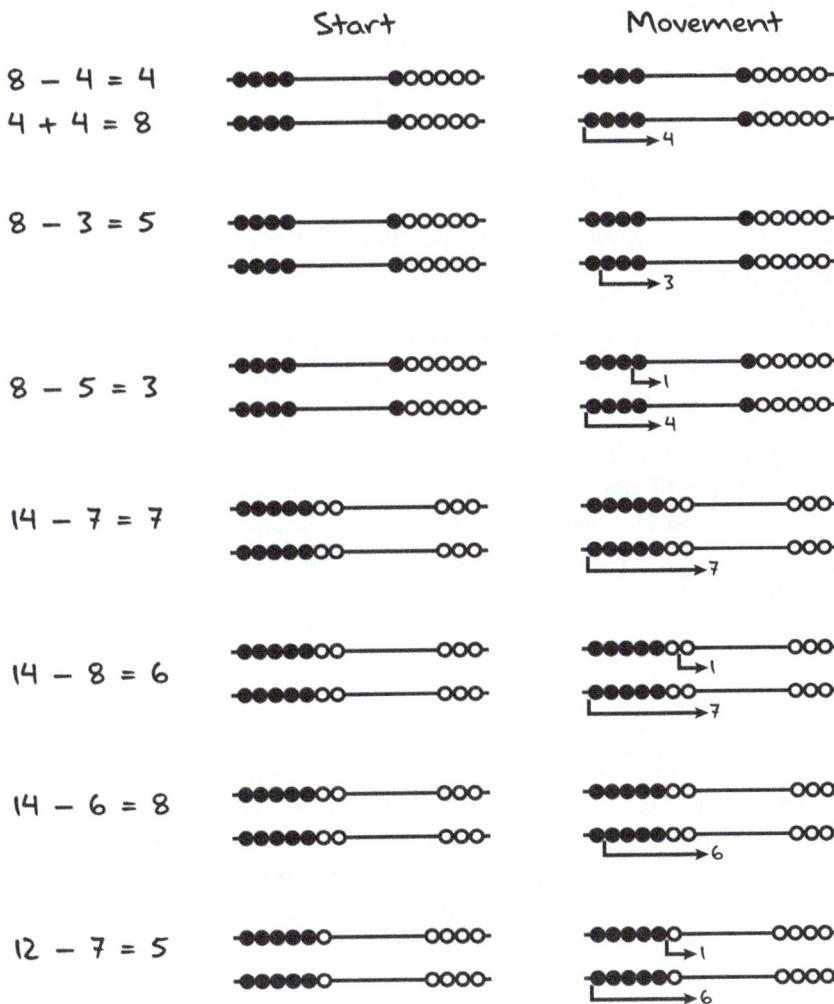

Chapter 5 • The Major Strategies for Subtraction Within 20   117

Here's a sample final display for this Problem String on an open number line:

Download a handy PDF for this Problem String.

https://qrs.ly/f5gl267

8 − 4 = 4
8 − 3 = 5
8 − 5 = 3

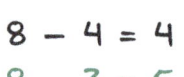

14 − 7 = 7
14 − 8 = 6
14 − 6 = 8

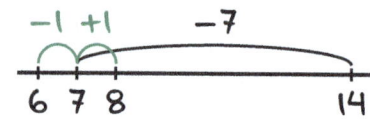

12 − 7 = 5

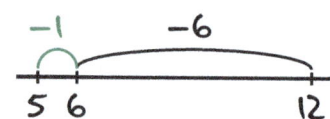

FREQUENTLY ASKED QUESTIONS

Q: What if my students still do not own very many doubles? Should I still do the tasks you recommend to develop Using Doubles to Subtract?

A: Make sure you have worked to build the Using Doubles to Add strategy with many experiences. Practice doubling with students (see Chapter 8). Also remember that these tasks that are designed to help students learn to use doubles also simultaneously reinforce the doubles.

Using Doubles to Subtract relies on students recognizing that the first number is double, or close to double, the second number in a subtraction problem. Continue to strengthen student comfort with doubles by doing doubles routines, giving random numbers, and asking for the double. A simple but useful routine to extend comfort with adding doubles toward recognizing doubles in a subtraction problem is to ask them to think about halving numbers.

Tasks where we purposefully plan the numbers can also help students develop this strategy:

Eliza had 12 colored pencils and gave away 6. How many colored pencils did she have left?

Oh wait, one more friend wanted a colored pencil, so she actually gave away 7 colored pencils. How many does she have now?

Benji found 14 marbles but then lost 7. How many marbles did he have?

Oh wait, he recounted, and he actually lost 8 marbles! How many does he have now?

> **TIP**
>
> If you're working on Using Doubles to Subtract and a student uses a different strategy to solve a problem, try to keep the conversation about reasoning and not about insisting they use a specific strategy. You can say, "Great way to remove to 10 and add then subtract the leftovers! I can understand your thinking. I wonder, do you understand what some of your classmates are doing? They are using 6 + 6 or 7 + 7 to help. Does that make sense to you?" Remember that it's not about the answer; it's about building multiple connections.

THE REMOVE 10 AND ADJUST STRATEGY

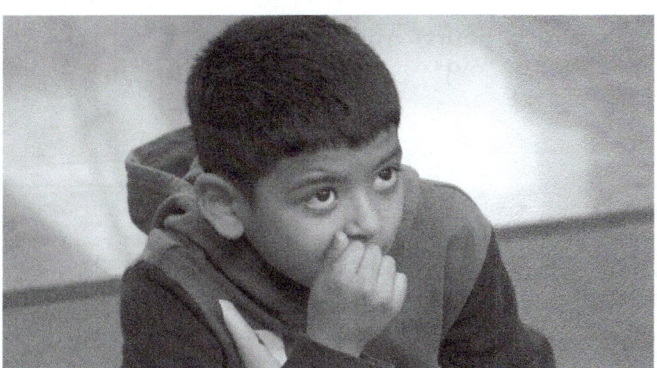

Single-digit Subtraction
Remove to 10
Using Doubles to Subtract
Remove 10 & Adjust
Find the Distance/Difference

Source: Adapted from Math Is Figure-Out-Able at https://www.mathisfigureoutable.com/ with CC Attribution-NoDerivatives 4.0 International License

The Remove 10 and Adjust Strategy requires students to think ahead, realizing they can remove a bit too much and then adjust. Students start by subtracting 10 from the minuend, then adjust by adding back the extra that was removed.

For a problem like 16 – 9, think 16 – 10 = 6. But we removed 1 too much, so now we need to add that 1 back to 6 to get 7. So, 16 – 9 = 7.

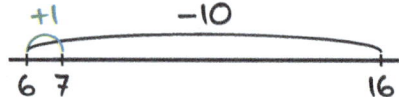

The Remove 10 and Adjust strategy uses and helps strengthen the idea of the teen numbers being 10 plus a single digit number. It also helps students grapple with the important inverse relationship of addition and subtraction. When you subtract 10 but only wanted to subtract 9, you need to add 1 more back to compensate for removing too much.

The Remove 10 and Adjust strategy is great for removing 8 or 9 from teen numbers. For example, 15 – 9 can be thought of as 15 – 10 and then adding back 1, 15 – 10 + 1 = 5 + 1 = 6. This strategy is most useful when a single-digit number close to 10 is being removed from a teen number, where the removal requires bridging 10. This strategy is not as useful for problems where the subtrahend is significantly smaller than 10 or when the subtrahend is smaller than the ones digit of the teen number. For example, removing 10 and then adding back 5 for 14 – 5 might be fun to play with, but it is not great for efficiency.

> ### FREQUENTLY ASKED QUESTIONS
>
> **Q:** This Remove 10 and Adjust strategy seems difficult for many of my students. They can't seem to remember which way they should adjust. Should I just tell them when they over subtract, they have to add back?
>
> **A:** Remember that the goal is to make sense of the relationships, not just get answers. Rather than telling students what to do, use models to help students realize what is happening when they subtract too much.

Part III • Developing Additive Reasoning

MODELS TO BUILD THE REMOVE 10 AND ADJUST STRATEGY

Use both a number rack and an open number line to represent student thinking for the Remove 10 and Adjust strategy. Use students' growing familiarity with the models to your advantage. As you represent the problems, invite students to make sense of the relationships using the models.

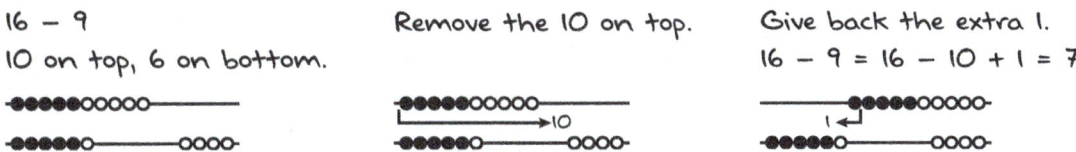

Transition to an open number line, putting tick marks where it is helpful to make the relationships within the strategy visible.

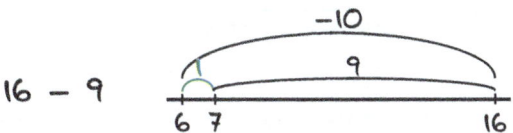

Note that with the modeling of this strategy, the –9 itself is not shown on the number line. Take time to discuss with students how the –9 is represented here even though there isn't a 9 in the model. Once or twice, draw the jump of 9 in during discussion to help students see the relationship between 10, 9, and 1.

HOW TO TEACH THE REMOVE 10 AND ADJUST STRATEGY

To help students both generalize and get better at removing the friendly 10 and adjusting, engage students in Problem Strings like the one in Table 5.3.

TABLE 5.3 • Problem String Using the Remove 10 and Adjust Strategy

PROBLEMS	TEACHER
14 – 10	"What is 14 – 10? How are you thinking about that? The teens are just a ten and some more, aren't they?"
14 – 9	"How about 14 – 9? Could you think about the last problem to help you?"

(Continued)

(Continued)

PROBLEMS	TEACHER
16 – 10	"A friendly ten again? Nice!"
16 – 9	"How does this relate to 16 – 10?"
15 – 10	*Repeat.*
15 – 8	"I wonder if there is anything up here that might help us. It seems helpful to remove a friendly number that is a little over when subtracting a trickier number."

Here's a sample final display for this Problem String on a number rack:

	Start	Movement	Result
14 – 10 = 4			
14 – 9 = 5 (14 – 10) + 1			
16 – 10 = 6			
16 – 9 = 7 (16 – 10) + 1			
15 – 10 = 5			
15 – 8 = 7 (15 – 10) + 2			

Here's a sample final display for this Problem String on an open number line:

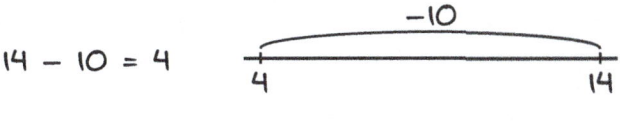

$14 - 10 = 4$

$14 - 9 = 5$
$(14 - 10) + 1$

$16 - 10 = 6$

$16 - 9 = 7$
$(16 - 10) + 1$

$15 - 10 = 5$

$15 - 8 = 7$
$(15 - 10) + 2$

Download a handy PDF for this Problem String.

https://qrs.ly/1rgl268

Tasks where we purposefully plan the numbers can also help students develop this strategy:

> Lucy has 13 cupcakes. She planned to give 10 to her classmates. How many would she have left to take home to her family?
>
> Lucy took the 13 cupcakes to school but only gave away 8 to her classmates. How many did she have left to take home to her family?
>
> Rahul had 17 toy cars and lost 10 of them. How many toy cars does he have left?
>
> Rahul realized that he actually only lost 9 of them. How many toy cars does he have?

TIP

Use contexts where students give money but realize they've given too much. Somehow students get motivated to figure out how much when money is involved.

TRY IT

Solve $14 - 8$ using the Remove 10 and Adjust strategy. Represent on a number rack and an open number line.

Remove 10 and Adjust is a handy strategy because you keep track as you go and end on the answer.

IMPLICATIONS OF THE REMOVE 10 AND ADJUST STRATEGY FOR DEVELOPING MATHEMATICAL REASONING

Students will continue to use an extended version of the Remove 10 and Adjust strategy within multi-digit subtraction problems. This later version is called the Remove a Friendly Number Over strategy (Chapter 7), where instead of always removing 10 and adjusting, you remove more general friendly numbers, often a multiple of 10, and adjust from there.

For example, to subtract 36 − 18, students can think about removing the friendly 20 first. 36 − (20 − 2) = 36 − 20 + 2 = 16 + 2 = 18.

The same reasoning applies for even bigger numbers, like 128 − 99: 128 − (100 − 1) = 128 − 100 + 1 = 28 + 1 = 29.

FREQUENTLY ASKED QUESTIONS

Q: What is a friendly number?

A: Great question! Friendly numbers are any number that it is notably easier to do math with than other numbers. What numbers are friendly can differ from person to person based on their experience, but as a starting point, multiples of 10 are usually the first to become friendly, followed by multiples of 5. For experienced reasoners, all kinds of numbers can end up friendly, including multiples of 25, 12, and 4.

FINDING THE DISTANCE/DIFFERENCE STRATEGY

Single-digit Subtraction
Remove to 10
Using Doubles to Subtract
Remove 10 & Adjust
Find the Distance/Difference

Source: Adapted from Math Is Figure-Out-Able at https://www.mathisfigureoutable.com/ with CC Attribution-NoDerivatives 4.0 International License

Subtraction has two meanings: difference/distance and removal. Each of the prior strategies is based on removal, where you start with the minuend and remove the subtrahend. But certain problem types beg for a distance interpretation. Similarly, certain numbers in relationships beg for a distance interpretation. Therefore, finding the Distance/Difference is an important strategy for subtraction.

When subtracting by finding the distance/difference between the numbers, students are thinking about the span between the two numbers. For example, the answer 14 − 8 can be found by thinking about how far apart 8 and 14 are on the number line.

$$14 - 8 = 6 \qquad \underset{8 \quad 10 \qquad 14}{\overset{2 \; + \; 4 \; = \; 6}{\frown}}$$

> Subtraction has two meanings: difference/distance and removal.

When students solve subtraction problems using the Distance/Difference Strategy, they often use their addition strategies. Therefore all of the information about those strategies applies here.

COMPARING THE SINGLE-DIGIT SUBTRACTION STRATEGIES

Each of these three strategies in Table 5.4 serves different purposes both in building additive thinking and preparing learners for other more sophisticated thinking later.

TABLE 5.4 • Comparing the Single-Digit Subtraction Strategies

STRATEGY	DEALING WITH THE SUBTRAHEND	THINKING ABOUT	WHAT IT BUILDS
Remove to 10	Break the subtrahend into the part to get to 10 and the leftovers	"Would getting back to 10 help me?"	Removing to 10 by a single-digit number (the meaning of the teen numbers), decomposing the other subtrahend, partners of 10, the primacy of 10
Using Doubles to Subtract	Remove half of the minuend and adjust to the subtrahend	"Can I adjust the double I know?"	Doubles and adjusting

(Continued)

(Continued)

STRATEGY	DEALING WITH THE SUBTRAHEND	THINKING ABOUT	WHAT IT BUILDS
Remove 10 and Adjust	Remove 10 using knowledge of a teen minus 10, adjust back the overage	*Requires planning ahead*, "Can I remove a bit too much and adjust back?"	Adding 10 to a single-digit number (meaning of teen numbers), partners of 10 (as they decide how much to adjust back), and counting backward
Finding the Distance/Difference	Find the distance between the subtrahend and the minuend	"Can I find how far apart these numbers are?"	There are two meanings of subtraction: removal and distance/difference.

> **TRY IT**
>
> Find 15 − 8 using each of the four strategies. Represent the strategy with at least one model.

Conclusion

Learning subtraction within 20 is important. Help students learn important additive relationships for subtraction by developing the major strategies. These relationships and strategies create the best foundation for success with multi-digit operations. As students connect what they are learning about subtraction to the major strategies in addition, they are strengthening their sense that subtraction is figure-out-able!

Discussion Questions

1. Which of the major subtraction strategies for numbers within 20 do you find yourself gravitating toward or tend to think of first? Which is less obvious to you?
2. Which of these three strategies involves using partners of 10? In what ways?

3. Enact with a thought partner: you're working with a student who does not just know the answer to 17 – 9. Choose from the following prompts and create a plausible back and forth focused on a pattern of questioning: "How do you want to think about 17 – 9? Do any relationships stand out to you when you think about 17 and 9? Could you think of 17 as 10 and 7? Does that help you think about removing 9? How do you want to remove that 9?"

4. Choose another often-missed subtraction problem, like 15 – 8. Write your own prompts, and create a plausible back-and-forth focusing on the pattern of questioning.

5. Interview your students to find which subtraction facts they know and which they do not know yet. Make a comprehensive list of the facts on most students' lists. Make a plan to work on those by giving students varied experiences with word problems, Problem Strings, games, and other routines.

CHAPTER 6

The Major Strategies for Double-Digit Addition Strategies

FIGURE 6.1 • The Second Level of Sophistication in Mathematical Reasoning

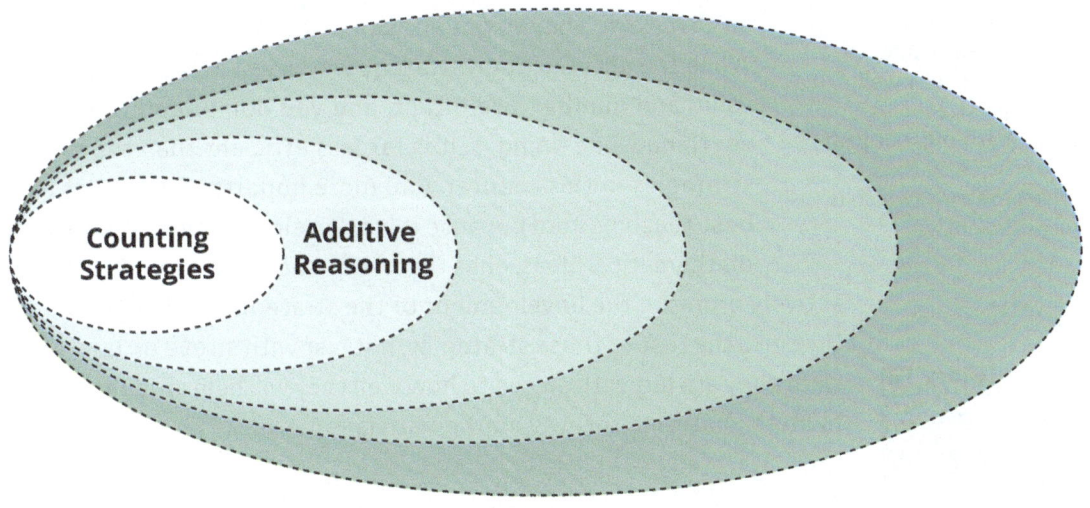

Source: Adapted from Math Is Figure-Out-Able at https://www.mathisfigureoutable.com/ with CC Attribution-NoDerivatives 4.0 International License

I was sitting in a ballroom at a national math teaching conference, listening to Scott Flansburg, who is known as the Human Calculator, talk about how he solves many-digit problems very quickly. He showed an example of adding numbers like 49 + 76 + 37, where you accumulate as you go, starting from the largest numbers.

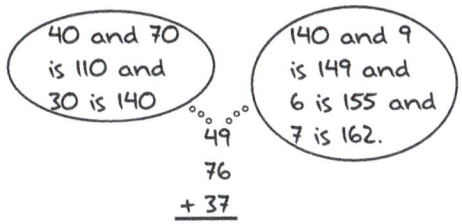

Since then, I've called this an accumulation strategy. Because you are accumulating the whole time, you end on the total sum. It's a handy strategy when you need to add several numbers or numbers with several digits, and you do not have a pencil to record thinking. He suggested that it maintains place value, is not a procedure to memorize because it just makes intuitive sense, and since you end with the answer, you don't need a pencil.

> ### TRY IT
>
> Solve 56 + 28 using this accumulation strategy. Notice how you end on the answer.

> **TIP**
>
> As you read the rest of this chapter, pay attention to how well the major strategies teach math-ing, not just how efficiently they solve problems (which is very efficiently).

Maybe you're reading this book and thinking, "Yes! That accumulation strategy is so doable. Let's teach all our students to do that. Bam. They've got answers. We win." Hang on a minute. While the accumulation strategy works well to keep the place value and number size intact, and you don't need to record anything while doing it, it is far less efficient than the major strategies in this chapter. And more importantly, it's not the best teaching tool because it won't help students learn other mathematical ideas, ones that we can develop in students if we actively promote the development of the strategies I am suggesting. Dive into the rest of these strategies, not just with an eye on how efficient they are but with an eye to how well they will help students learn to math, math-ing the way mathematicians math.

See these students reasoning in action.

https://qrs.ly/sjgl26b

Stephanie Lugo, a second-grade teacher, uses Problem Strings to develop the Get to a Friendly Number strategy with her second-grade class.

By the end of the Problem String, students are solving 37 + 14 by Getting to a Friendly Number.

DEVELOPING MULTI-DIGIT ADDITION STRATEGIES

*Get to 10
 *Using Doubles to Add
  *Add 10 & Adjust
   *Remove to 10
    *Using Doubles to Subtract
     *Remove 10 & Adjust
      *Find the Distance/Difference
       Split by Place Value
        Add a Friendly Number
         Get to a Friendly Number
          Add a Friendly Number Over
           Remove a Friendly Number
            Remove to a Friendly Number
             Remove a Friendly Number Over
              Find the Distance/Difference
               Give and Take
                Constant Distance

Additive Strategies

*Within 20

Source: Adapted from Math Is Figure-Out-Able at https://www.mathisfigureoutable.com/ with CC Attribution-NoDerivatives 4.0 International License

When students have developed and are using the major relationships for single-digit addition and subtraction, it's time to develop addition and subtraction with multi-digit numbers.

Chapter 6 • The Major Strategies for Double-Digit Addition Strategies   131

For adding multi-digit numbers, the goal is developing a few major mathematical relationships that lead to students naturally and intuitively using a few important strategies. It's not about having students memorize and mimic a new or different set of steps. If they couldn't do that successfully with one algorithm, why would we want to now have them try to do the same failed thing with more than one procedure? No, it's about developing specific, important mental connections that make the strategies become natural outcomes—intuitive choices based on well-traveled logical mental paths. These paths are well-traveled not because students have put them to music or memorized a rhyme or a story. Rather, students have noticed patterns, tried using those patterns, shifted and refined that use, and made generalizations based on their experience. They have experienced having a teacher make those thoughts/relationships/connections/patterns visible (when my brain does that, it can look like that) and refined the words and visual images to fit old and new situations.

There are five major strategies to actively/purposefully help students develop for multi-digit addition. Those major strategies in Kindergarten through Grade 2 are:

- Splitting by Place-Value
- Add a Friendly Number
- Get to a Friendly Number
- Add a Friendly Number Over
- Give and Take

Each of these five strategies serves different purposes both in building Additive Reasoning and preparing learners for other more sophisticated thinking later. The overarching goal is for students to develop the relationships for all five of these strategies, even though students will probably learn to rely primarily on three of them. Splitting by Place Value is ideal when there's no regrouping, and Adding a Friendly Number Over or Give and Take when there is regrouping. But we need to develop all the major strategies. All of them together, even if not actively used individually, provide the best possible position to engage in further mathematical learning in subtraction, multiplication, and beyond.

FREQUENTLY ASKED QUESTIONS

Q: What do I do with students who already use some of these addition strategies?

A: You will have students for whom some of these strategies are naturally occurring. Allow students to use the strategies that come naturally even if you haven't yet introduced those strategies.

Q: But wait, the accumulation strategy isn't on your list of major strategies?

A: Correct—and neither is the swapping strategy. There are other fine strategies, but those listed here are the *major* strategies—the ones students need to solve any problem that's reasonable to solve without a calculator. These strategies build the relationships necessary for students to continue to progress. The accumulation strategy is cool, but we have to be laser focused on advancing the math for our students and teach what they need most.

Q: But, Pam, my students struggle learning the one algorithm. How can I expect them to now memorize five strategies?

A: The goal isn't memorizing and mimicking these five strategies. We know memorization and mimicry don't work! The goal is about helping students develop the major relationships underneath these strategies, learning to reason additively about bigger numbers. Constructing these relationships and using the strategies are mutually reinforcing.

THE SPLITTING BY PLACE VALUE STRATEGY

Addition

Split by Place Value
Add a Friendly Number
Get to a Friendly Number
Add a Friendly Number Over
Give and Take

Source: Adapted from Math Is Figure-Out-Able at https://www.mathisfigureoutable.com/ with CC Attribution-NoDerivatives 4.0 International License

Splitting by Place Value is the bedrock, beginning strategy for multi-digit addition. This strategy is often developed independently by students because it is based on students' developing a sense of place value. At first, students might be thinking about the number 38 as "the words you say after 35, 36, 37. That's 38." As students begin to grapple with a problem like 38 + 27, they think about *30 and some more* and *20 and some more*. They begin to make sense that 38 means 30 and 8. So, when students are adding something that they are starting to understand as 30 and 8 and 20 and 7, they think, "Well, let's just put those big numbers together first, 30 + 20, to get 50. Now, put the extra parts together, 8 and 7 is 15. And then I'll bring those together, 50 and 15, to get 65."

Because students add the tens together, their answer is already fairly reasonable: adding 20 (and extra) plus 30 (and extra) is going to be 50 (and extra). Bam—already close. Also, this grappling with 20 + 30 gives students experience with those big numbers: how 20 relates to 2 tens, 30 relates to 3 tens and simultaneously that 20 + 30 is 50 and 2 tens + 3 tens is 5 tens. They don't abandon the magnitudes by only thinking about the digits 2 + 3.

All this naturally germinating reasoning is directly undercut by digit-focused algorithms (Harris, 2025).

The Splitting by Place Value strategy can be an efficient strategy when the sum of the place values stays within the place at hand, where there's no need to carry over any part of the sum. For example, 52 + 37 is an excellent problem to Split by Place Value because you can just add the place values left to right. Notice that when people do this naturally (if they have not been force-fed an algorithm), they think about the big numbers first: 50 and 30 is 80, plus that 2 and 7 is 9, so 89. This is incredibly convenient for larger sums where there is no more than 9 of a particular unit, like 473 + 325. You can think about how 400 + 300 (700), plus the 70 + 20 (90), plus the 3 + 5 (8), is simply 798.

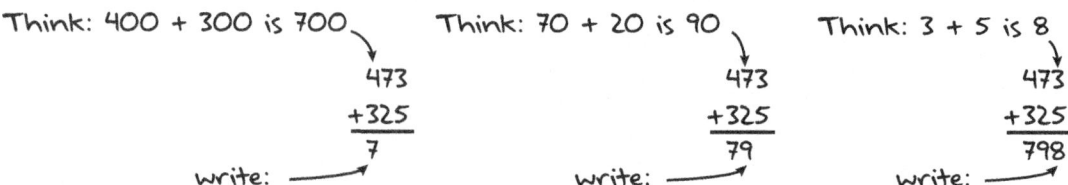

Splitting by Place Value focuses on the parts of both addends, the hundreds and tens and ones, instead of other strategies that keep addends whole and consider their magnitudes. For instance, in using an Over strategy with 473 + 325, the 473 is almost 475. And 475 plus 325 is 800, but that's just 2 too much for 473 + 325, so 798.

Most importantly, if students develop only this strategy, they may never develop many other relationships they need to develop. It's a necessary starting place but cannot be the finish line.

Splitting by Place Value is essential but not sufficient.

MODELS TO BUILD THE SPLITTING BY PLACE VALUE STRATEGY

To represent this Splitting by Place Value strategy, start with a splitting model and transition to equation models (Table 6.1).

TABLE 6.1 • Models Representing the Splitting by Place Value Strategy

	49 + 35	
SPLITTING MODEL	**VERTICAL EQUATION MODEL**	**HORIZONTAL EQUATION MODEL**
49 + 35 ╱╲ ╱╲ 40 9 30 5 ⌣ ⌣ 70 + 14 = 84	$49 = 40 + 9$ $+35 = 30 + 5$ $\overline{70 + 14 = 84}$	$49 + 35 = (40 + 9) + (30 + 5)$ $= (40 + 30) + (9 + 5)$ $= 70 + 14$ $= 84$

It is noteworthy that we do not use an open number line. The continuous open number line model does not lend itself to representing breaking up both *addends.*

HOW TO TEACH THE SPLITTING BY PLACE VALUE STRATEGY

By representing student thinking with appropriate models and using tasks and Problem Strings, you can help students develop the Splitting by Place Value strategy and the accompanying relationships. Because many students develop this strategy naturally on their own, you may not need to spend very much time here. The best thing you can do is *not* teach a step-by-step procedure that starts with the smallest numbers and demands that students focus on columns of digits.

Instead, give students problems where they are finding the sum of two-digit numbers. Ask them how they are thinking about adding those numbers together, and as they explain splitting up the numbers by place value, represent their thinking to make it visible.

Start with two digits that don't require regrouping, like 32 + 26. As they get good at splitting by place value, you can give them more complicated problems where there is a carryover from the ones place. This gets everybody better at Splitting by Place Value and introduces problems that will be ideal for the next strategies.

To help students both generalize and get better at splitting by place value, engage students in Problem Strings like the one in Table 6.2. This string is one that you would do after students have already done similar strings where there was no regrouping required.

TABLE 6.2 • Problem String Using the Splitting by Place Value Strategy

PROBLEM	TEACHER
30 + 20	"What's 30 and 20? How do you know?" *Represent with an equation.*
8 + 7	*Repeat.*
38 + 27	"What's 38 and 27? Did anyone add 30 and 20? Tell us about that." *Represent with a splitting model.*
50 + 10	*Repeat.*
4 + 9	*Repeat. Quickly.*
54 + 19	*Repeat.* "Seems like thinking about these numbers in pieces is helpful."
40 + 30	*Repeat.*
46 + 38	"I wonder if anyone made a helper problem for this one? Why 6 + 8? How was that helpful?"

Download a handy PDF for this Problem String.

https://qrs.ly/8ggl26i

Here's a sample final display for this Problem String:

30 + 20
8 + 7
38 + 27
50 + 10
4 + 9
54 + 19
40 + 30
46 + 38

30 + 20 = 50
8 + 7 = 15
38 + 27
 ∧   ∧
30 8 20 7
   ⤫
50 + 15 = 65

50 + 10 = 60
4 + 9 = 13
54 + 19
 ∧   ∧
50 4 10 9
   ⤫
60 + 13 = 73

40 + 30 = 70
46 + 38
 ∧   ∧
40 6 30 8
   ⤫
70 + 14 = 84

> **TRY IT**
>
> Solve 56 + 28 using a Split by Place Value strategy. Represent your strategy with a splitting model and an equation model, both horizontal and vertical.

IMPLICATIONS OF THE SPLITTING BY PLACE VALUE STRATEGY FOR DEVELOPING MATHEMATICAL REASONING

The implications for higher math is that Place Value Partial Sums is not enough. Too many well-meaning elementary textbooks have suggested that Place Value Partial Sums is enough—if students can split by place value, you can stop there. Students are reasoning with place value, getting correct answers, not following a script, what else do you need?

It turns out we need a lot more. Students need to learn to keep one addend whole and then decompose the other addend in significant ways. Not because we want to lock students into memorizing many procedures. Also not *just* because students need to start keeping one addend whole and decomposing the other. The more sophisticated strategies *are* important because they help students develop more sophisticated place value and magnitude relationships than you can get with just Splitting by Place Value.

> **TIP**
>
> When you are saying the numbers as you represent student thinking, emphasize the tens in the two-digit number to highlight the tens place and ones place of the number. For example, when you say 48, put the emphasis on forty in *forty*-eight—so as you write 48 = 40 + 8, you say, "*Forty*-eight is 40 and 8."

Developing the rest of the major strategies also gives us a chance to move to the open number line model, developing a sense of space with nearness, neighborhood, order, size, and measurement. Open number lines provide the opportunity for students to not only gain intuition that 49 is one before 50 but also that it's farther from 49 to 100 than 0 to 49. Students also visualize that there is a span between 49 and 100 that we can envision as distance traveled, not discrete numbers to count. On an open number line, we are looking at the distance between numbers, not counting the tick marks.

Understanding the span of number relationships represents significant conceptual growth for students because earlier counting was all about counting discrete objects. Now addition is not just finding the total number of discrete objects, but it can *also* be finding the total distance traveled, a measurement idea. While developing that sense, students are necessarily developing a linear map in their heads about the relationships between and among numbers.

> **TIP**
> Addition is not just finding the total number of discrete objects, but it can *also* be finding the total distance traveled, a measurement idea.

Splitting by Place Value is also not sufficient because in older grades students will need to add larger and more complex numbers, at which point splitting by place value gets too cumbersome and inefficient.

Students who are working to understand such problems will briefly entertain splitting by place value as they make sense of new and more complex numbers, but if they have developed the major strategies with smaller whole numbers, they will quickly switch to other, far more efficient strategies.

$$\begin{array}{r} 765.98 = 700 + 60 + 5 + 0.9 + 0.08 \\ +98.6 = \phantom{700 + {}} 90 + 8 + 0.6 \\ \hline 700 + 150 + 13 + 1.5 + 0.08 = 864.58 \end{array}$$

You might look at this and think, "That makes sense! We should do that. Students can make sense of the numbers!" You betcha! Stay tuned for even more sense making to come.

Alternately, you may be thinking, "That's so many pieces. All. The. Pieces." Ironically, it's the same number of pieces as the traditional algorithm, albeit with place value still intact. But either way, the aim is to get more sophisticated and efficient!

The Split by Place Value strategy is often referred to imprecisely in textbooks simply as "partial sums." "Partial sums" is a general description of breaking the problem into smaller addition problems and then adding them together. The strategy we are referring to here is a *specific* type of partial sum, where the numbers are split by place value and then those like-places are brought together. Therefore, we do not label this as the generic partial sums, but the *Place Value* Partial Sum strategy or, as we often call it, Splitting by Place Value.

THE NEXT TWO MAJOR STRATEGIES

Addition
Split by Place Value
Add a Friendly Number
Get to a Friendly Number
Add a Friendly Number Over
Give and Take

Source: Adapted from Math Is Figure-Out-Able at https://www.mathisfigureoutable.com/ with CC Attribution-NoDerivatives 4.0 International License

The next two strategies, Add a Friendly Number and Get to a Friendly Number, are on par with each other in sophistication, and students need experience to develop both. You can work on both at the same time and in either order. Later, compare the two to help students get better at each.

These two strategies share the feature that you keep one addend whole and add pieces of the other addend in sensible ways. How you break up the second addend is the difference between Get to a Friendly Number (where you break up the addend to reach the next friendly number first) and Add a Friendly Number (where you break up the addend to add a friendly part first). It's all about the solver's first move.

Let's take a look at these two important strategies.

THE ADD A FRIENDLY NUMBER STRATEGY

Addition
Split by Place Value
Add a Friendly Number
Get to a Friendly Number
Add a Friendly Number Over
Give and Take

The Add a Friendly Number Strategy means keeping one addend whole and first adding a friendly part of the second addend, then adding the rest. As students realize that they can use the patterns in our number system to think about adding friendly numbers, this strategy becomes a natural inclination.

For example, young students often start to realize that they can think about any number plus 10 (like 34 + 10) as 30 and one more group of 10, and that extra 4. If they encounter a problem like 34 + 17, they might think, "I don't know 34 and 17, but I do know 34 and 10. I'll do that and then just add the rest. So, 34 + 10 is 44, so now I'll just add the leftover 7."

Similarly, a student might think about a problem like 46 + 38, "I know 46 and 30, that's like 40 and 30 with the 6, so 76. Now I'll just keep adding that extra 8."

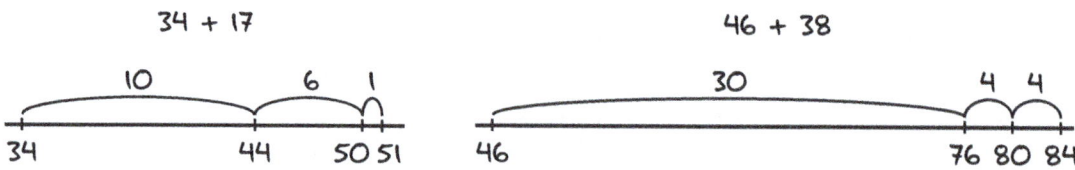

TIP

Note how in each of these examples the Add a Friendly Number strategy takes fewer steps than Splitting by Place Value. That is one payoff of increased sophistication.

Note that a goal is to help students add a *large* friendly number. For 46 + 38, students could keep 46 whole and then add 3 jumps of 10 and then the 8. Work with students to keep 46 whole and add the *larger* friendly 30 in one jump and then the 8. The goal is *fewer, bigger* jumps.

In general, the Add a Friendly Number strategy looks like the following:

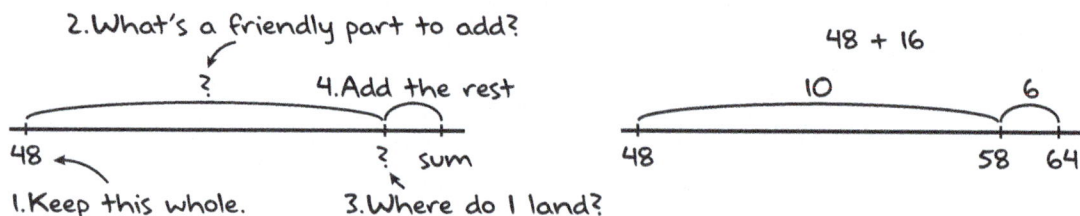

MODELS TO BUILD THE ADD A FRIENDLY NUMBER STRATEGY

You may find it helpful at first to represent student thinking using a splitting model, to build off how students have been seeing their thinking when they've been splitting by place value

(Table 6.3). Quickly move to the open number line as the model of choice as it lends itself to more strategies and longevity.

TABLE 6.3 • Representation of Student Thinking Using the Splitting Model

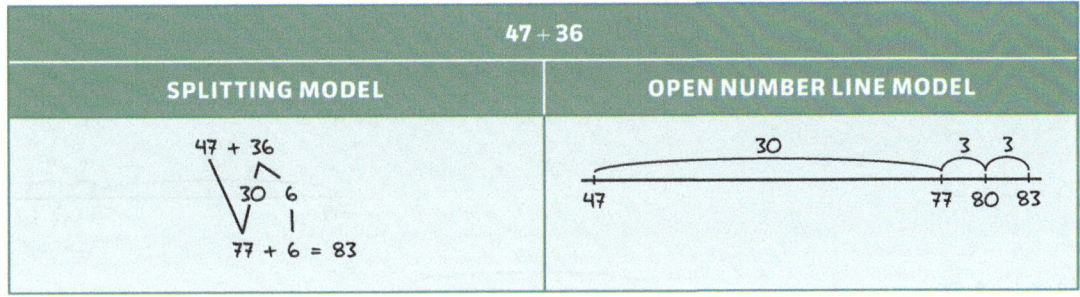

SPLITTING MODEL	OPEN NUMBER LINE MODEL

HOW TO TEACH THE ADD A FRIENDLY NUMBER STRATEGY

To help students both generalize and get better at adding a friendly number, engage students in Problem Strings like the one in Table 6.4.

TABLE 6.4 • Problem String Using the Add a Friendly Number Strategy

PROBLEM	TEACHER
27 + 10	"What is 27 plus 10?" *Represent on an open number line.*
27 + 14	*Repeat.* "Did anyone use the problem before? Could you? How?"
49 + 10	*Repeat.*
49 + 13	*Repeat.* "Did anyone use the problem before? How?"
36 + 10	*Repeat. Quickly.*
36 + 20	*Repeat.* "Did anyone make two jumps of 10? One jump of 20?"
36 + 25	"How could you use the problem before to help? It seems helpful to add a friendly part and then add the rest."
26 + 17	"Could you make up your own helper for this one, something friendly? Sure seems helpful to add a friendly part and then the rest."

Download a handy PDF for this Problem String.

https://qrs.ly/c7gl26p

Here's a sample final display for this Problem String:

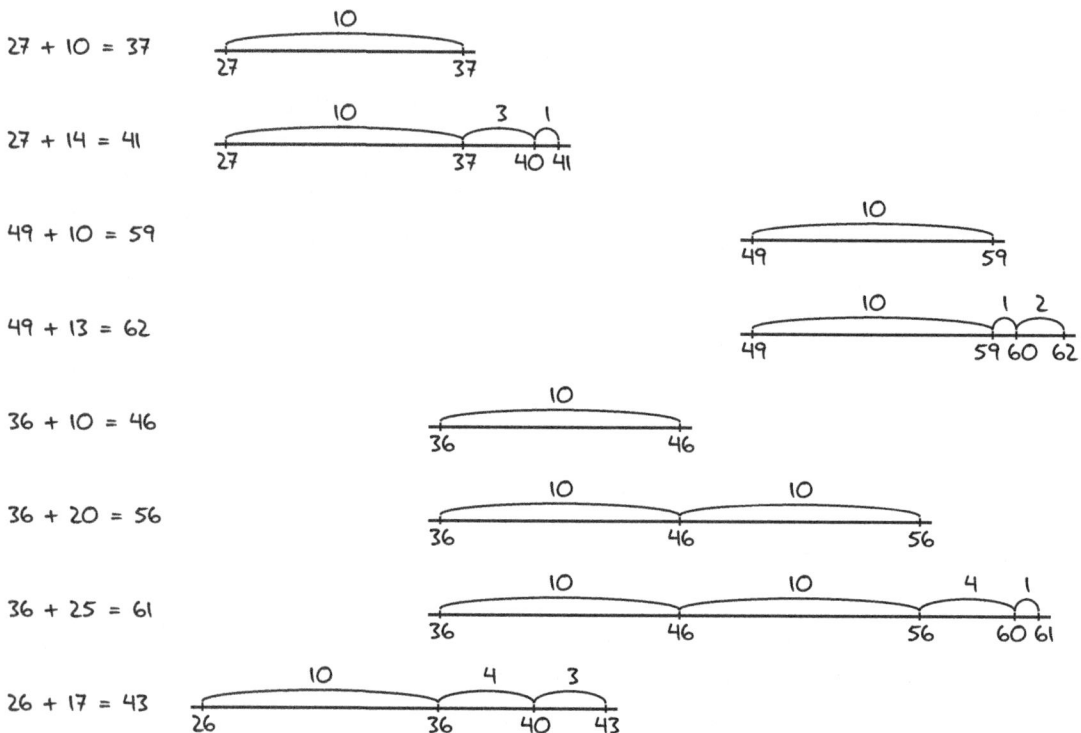

Tasks where we purposefully plan the numbers can also help students develop this strategy.

> The second grade ordered 38 cheese and 20 pepperoni pizzas. How many total pizzas did they order?
>
> Actually, they forgot some orders. They need 38 cheese and 24 pepperoni. How many did they really order? Could you use the first order to help? How?
>
> They ordered 37 sodas and 10 lemonades. How many drinks?
>
> Actually, they need to order 37 sodas and 16 lemonades. How many total drinks do they actually need? Could you use the first order to help? How?

Count Arounds, as described in Chapter 8, are a great way to help students realize that they can add multiples of 10 to any number. Start with a single-digit number and add 10 each time. Line up the numbers so that students have opportunities to notice, describe, and generalize place-value patterns.

FREQUENTLY ASKED QUESTIONS

Q: What if my students do not realize what a friendly number is? Should I define that first so they know what they should add?

A: Rather than define a friendly number, do the tasks suggested here to help students realize which numbers are friendly in these problems. Notice that in the Problem String, the pizza problems, and the suggested Count Around students are experiencing friendly numbers. As they do, create class conversations about what makes those numbers friendly.

TRY IT

Solve 56 + 28 using an Add a Friendly Number strategy. Represent your strategy with both a splitting model and a number line.

IMPLICATIONS OF THE ADD A FRIENDLY NUMBER STRATEGY FOR DEVELOPING MATHEMATICAL REASONING

While Add a Friendly Number is an important early strategy, is becomes tedious the more digits that are involved. Once students have added a friendly number, there is often more work to do to cross over the next place value, which becomes cumbersome. Notice that in the example that follows, the student grapples with adding the leftover 79 to 589, and the process involves many little jumps. After students have developed this strategy, the goal is to make bigger, fewer jumps.

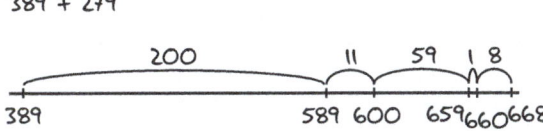

Start with 389, add 200, then add 70. I don't know 589 + 70 so I'll add 11 to get to 600, how much left? Um, 70 − 11 = 59, so add 59. Now add 9, by adding 1 to 660, then 8 more.

Therefore, Adding a Friendly Number is less of a strategy we want added to a student's long-term toolbox and more a way to help students develop and strengthen important relationships. It is often more efficient than Splitting by Place Value, but its true value lies in preparing students for even more sophisticated reasoning. We want to promote the development of this strategy because it strengthens other essential concepts by

- helping students make sense of where numbers fit with each other: to add 36 and 28, keeping 36 whole, students must think about 36 as a starting point, where 36 lives, and what's around it; and
- helping students generalize the notions of place value where they consider *any number plus 10* or *any number plus a multiple of 10*. Examples include 37 + 10, 49 + 20, 67 + 30.

Later, students will generalize *any number plus* 100, 1000, or even decimals and fractions plus whole numbers, like the following:

- 574 + 200 to find 574 + 240
- 4387 + 2000 to find 4387 + 2024
- 2.8 + 1 to find 2.8 + 1.3
- $4\frac{4}{5} + 3$ to find $4\frac{4}{5} + 3\frac{1}{2}$.

THE GET TO A FRIENDLY NUMBER STRATEGY

Addition
Split by Place Value
Add a Friendly Number
Get to a Friendly Number
Add a Friendly Number Over
Give and Take

Source: Adapted from Math Is Figure-Out-Able at https://www.mathisfigureoutable.com/ with CC Attribution-NoDerivatives 4.0 International License

The Get to a Friendly Number strategy entails keeping one addend whole, then adding a portion of the second addend to get to the next friendly number, then adding the rest. As students decide to keep one addend whole, they consider partners to friendly numbers and then add the leftover large amounts after that.

For example, to add 38 + 17, a student realizes that 38 is close to 40, so they add 38 + 2 to get there. They then determine what's left to add (17 − 2 = 15) and add the leftover 15 to 40, landing on 55.

Similarly, a student adding 46 + 28 thinks about getting from 46 to 50 by adding 4. Since they were supposed to add 28, they need to add 24 more, so 50 + 24 = 74.

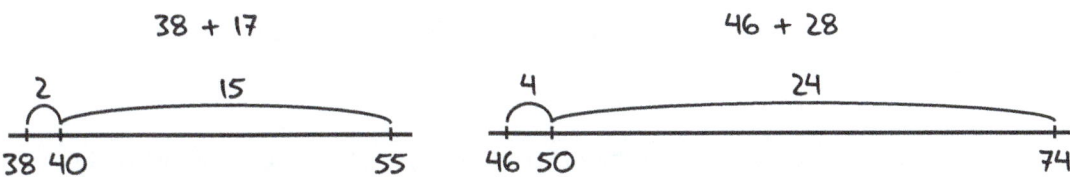

In the Get to a Friendly Number strategy, students are grappling with finding and recognizing the next friendly number. Consider 57 + 35. Starting from 57, what is the next friendly number? Is it 58, 60, 75, 100? It depends on the numbers in play! What is friendly that will help in *this* problem?

This helps the student strengthen their sense of what friendly numbers are, using partners of 10 to get to that friendly number and partitioning the other addend to determine how much is left to add. These build a sense of place value and magnitude, the size of numbers, and how they relate.

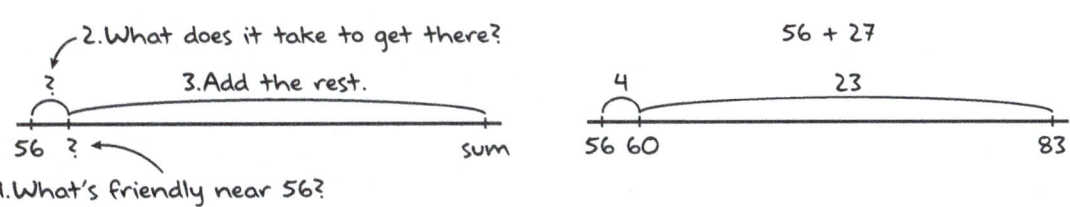

MODELS TO BUILD THE GET TO A FRIENDLY NUMBER STRATEGY

Just like with the Add a Friendly Number strategy you may find it helpful to briefly represent student thinking using a splitting model (Table 6.5). Quickly move to the open number line as it lends itself to more strategies and longevity.

TABLE 6.5

	47 + 35	
SPLITTING MODEL	**OPEN NUMBER LINE MODEL**	
47 + 35 → 3, 32; 50 + 32 = 82	47 →(3)→ 50 →(32)→ 82	

HOW TO TEACH THE GET TO A FRIENDLY NUMBER STRATEGY

To help students both generalize and get better at getting to a friendly number, engage students in Problem Strings like the one in in Table 6.6.

TABLE 6.6 • Problem String Using the Get to a Friendly Number Strategy

Download a handy PDF for this Problem String.

https://qrs.ly/c7gl26p

PROBLEM	TEACHER
28 + 2	"What is 28 and 2?" *Represent quickly on an open number line.*
28 + 7	*Repeat.* "Did anyone use the problem before? Could you? How?"
44 + 6	*Quickly.* "Partners of 10 sure are handy."
44 + 18	"Did anyone use the problem before? How? And how did you add 50 and 12? Anyone add it all in one jump?"
37 + 3	*Quickly.*
37 + 28	"How did you use the problem before? How much was left after getting to 40? Did anyone add that leftover 25 in one jump?"
26 + 17	"Could you make up your own helper for this problem? Why 26 + 4? What is it about 30 that makes it friendly? Sure seems like adding to a friendly number is helpful."

Here's a sample final display for this Problem String:

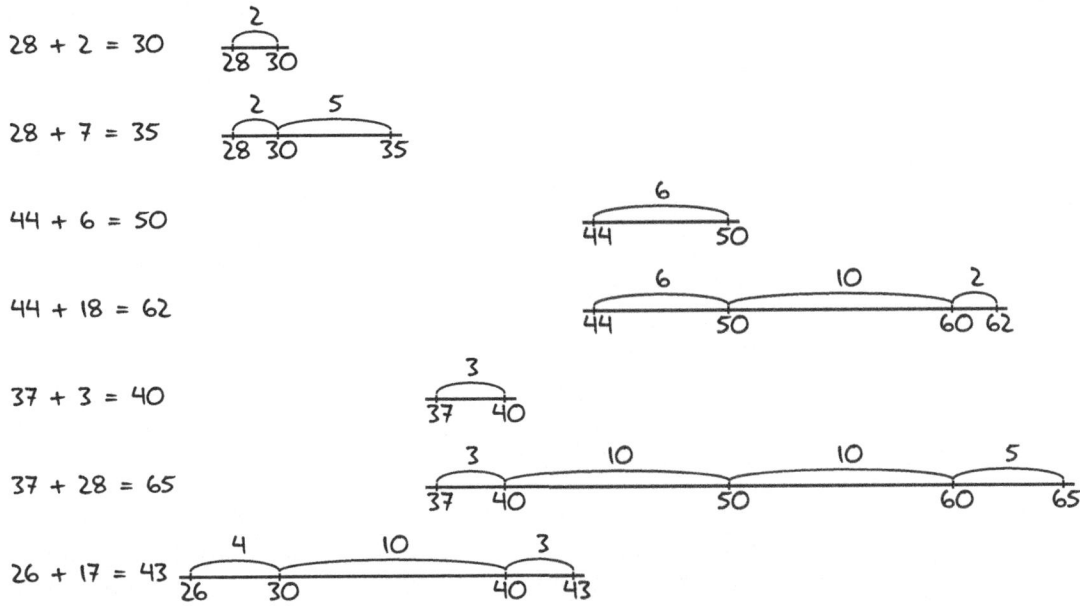

For this strategy to work well, students need to own partners of 10, and partners of 10 within multiples of 10. Notice that to think about 26 + 17 using Get to a Friendly Number, the student needs to recognize that 26 is close to a friendly number and that friendly number is 30. Using what they know about 6 and 4 being partners of 10, they can extend that partnership to know that 26 and 4 is the next 10, which is 30.

Tasks where we purposefully plan the numbers can also help students develop this strategy.

> The second grade ordered 38 cheese and 2 pepperoni pizzas. How many total pizzas did they order?
>
> Oh wait, they missed that the order form actually said 38 cheese and 16 pepperoni. How many did they order? Could you use the first order to help? How?
>
> They ordered 47 sodas and 3 lemonades. How many drinks?
>
> Oops, they need to order 47 sodas and 25 lemonades. How many total drinks do they need? Could you use the first order to help? How?

TIP
I Have, You Need strikes again. It's not about rote-memorizing partners of 10—it's about figuring them often, so those partners become intuitive. When students see 26, one of the things they think about is that 26 is 4 from 30.

Chapter 6 • The Major Strategies for Double-Digit Addition Strategies

> **TRY IT**
>
> Solve 56 + 28 using a Get to a Friendly Number strategy. Represent your strategy with a splitting model and a number line.

IMPLICATIONS OF THE GET TO A FRIENDLY NUMBER STRATEGY FOR DEVELOPING MATHEMATICAL REASONING

The Get to a Friendly Number strategy is often more efficient than the Add a Friendly Number strategy, especially when students realize and get good at adding the rest of the decomposed addend in one large jump.

TIP

If students are getting to a friendly number and then start to add the rest in many little jumps, ask them, "Do you know 60 and 32? Could you add the 32 in one big jump?" Even if they cannot *yet*, they've heard the suggestion that it's something to try for.

When students first start developing the Get to a Friendly Number strategy, they tend to make many small jumps. They think about 57 + 35 by starting at 57 and adding 3 to get to 60. Then they think about adding the rest in pieces that make sense to them, often adding 10 at a time and then the leftover 2.

You can help students begin to make fewer, bigger jumps by reminding them that they have been adding 30 in one group of 30 during Add a Friendly Number strings and by sharing students' thinking who reason to add 32 in one jump. Make the thinking visible and line up the number lines so that the smaller jumps of 10 and 2 are enveloped by the big jump of 32.

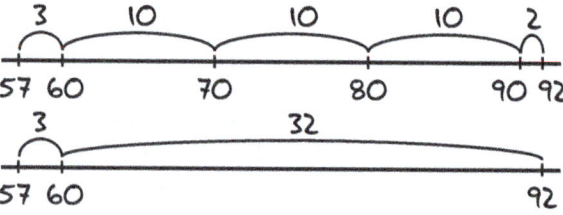

57 + 35

Like the Add a Friendly Number strategy, this Get to a Friendly Number strategy is more *sequential* in nature. That means that students make moves in sequence—start by making a move that makes sense, do the next move that makes sense, and so on, in a sequence of decisions. As students get better at getting to a friendly number and then adding the rest in one large jump, they are preparing for the most sophisticated addition strategy, Give and Take. Give and Take is a more *simultaneous* strategy, where students consider the outcome of their moves in their choices.

THE ADD A FRIENDLY NUMBER OVER STRATEGY

Addition
Split by Place Value
Add a Friendly Number
Get to a Friendly Number
Add a Friendly Number Over
Give and Take

Source: Adapted from Math Is Figure-Out-Able at https://www.mathisfigureoutable.com/ with CC Attribution-NoDerivatives 4.0 International License

The Add a Friendly Number Over strategy is a close relative to the Add a Friendly Number strategy. In this version, students think ahead, add a friendly number that is a bit too much, and then adjust back as needed. Just like with Add a Friendly Number, students keep one addend whole, but when they think about adding the second addend, they consider friendly numbers more than the second addend, knowing it would be efficient to add something too big, because then they can subtract off the extra.

For a problem like 38 + 19, think 38 + 20 = 58. We added 1 too much, so now we need to back up 1 from 58 to 57. So, 38 + 19 is 57.

Add a Friendly Number works well for problems where one of the addends is close to a friendly number, like 49, 28, 57. Examples include 48 + 29, 45 + 38 and even 56 + 99 (friendly 100, anyone?).

In general, the Add a Friendly Number Over strategy can look like the following:

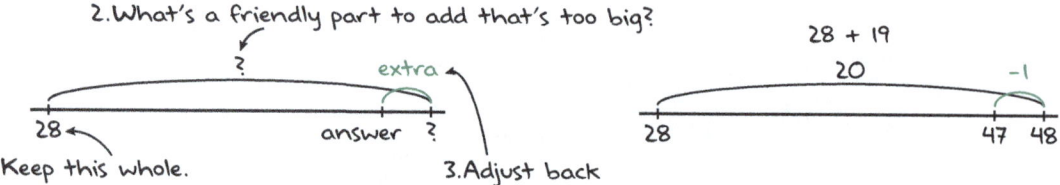

> **TIP**
>
> Use open number lines as your main model to represent student thinking for the Add a Friendly Number Over strategy. I've shown you the equation model here for your reference, but do not bog students down with equations too soon.

This strategy is also helpful to build some subtraction sense because it gets students moving up *and* down the number line, forward and back. Students are thinking about what comes after and before numbers, which is part of *number sense*.

MODELS TO BUILD THE ADD A FRIENDLY NUMBER OVER STRATEGY

Use both open number lines and equations to represent the Add a Friendly Number Over strategy (Table 6.7).

TABLE 6.7 • Number Lines and Equations Representing the Add a Friendly Number Over Strategy

47 + 38	
OPEN NUMBER LINE MODEL	**EQUATION MODEL**
(number line showing 47 → 85 with +40, then -2 back to 87... 85 87)	47 + 38 = 47 + (40 − 2) = (47 + 40) − 2 = 87 − 2 = 85

HOW TO TEACH THE ADD A FRIENDLY NUMBER OVER STRATEGY

To help students both generalize and get better at adding a friendly number that is too big and adjusting, engage students in Problem Strings like the one in Table 6.8.

TABLE 6.8 ● Problem String Using the Add a Friendly Number Over Strategy

PROBLEM	TEACHER
28 + 10	"What is 28 and 10?" *Represent quickly on an open number line.*
28 + 9	*Repeat.* "Did anyone use the problem before? Could you? How?"
44 + 10	*Quickly.* "We've been working on adding 10 to any number, haven't we?"
44 + 9	"Did anyone use the problem before? How? Adding a bit too much and then adjusting?"
37 + 20	*Elicit and represent both strategies of 2 jumps of 10 and 1 jump of 20.*
37 + 19	"How did you use the problem before? How is 19 related to 20? Can you put words to why you are subtracting in an addition problem?"
26 + 19	"Could you make up your own helper for this problem? Why 26 + 20? Sure seems like it's helpful to add a bit too much and then adjust."

Download a handy PDF for this Problem String.

https://qrs.ly/ncgl26r

Here's a sample final display for this Problem String:

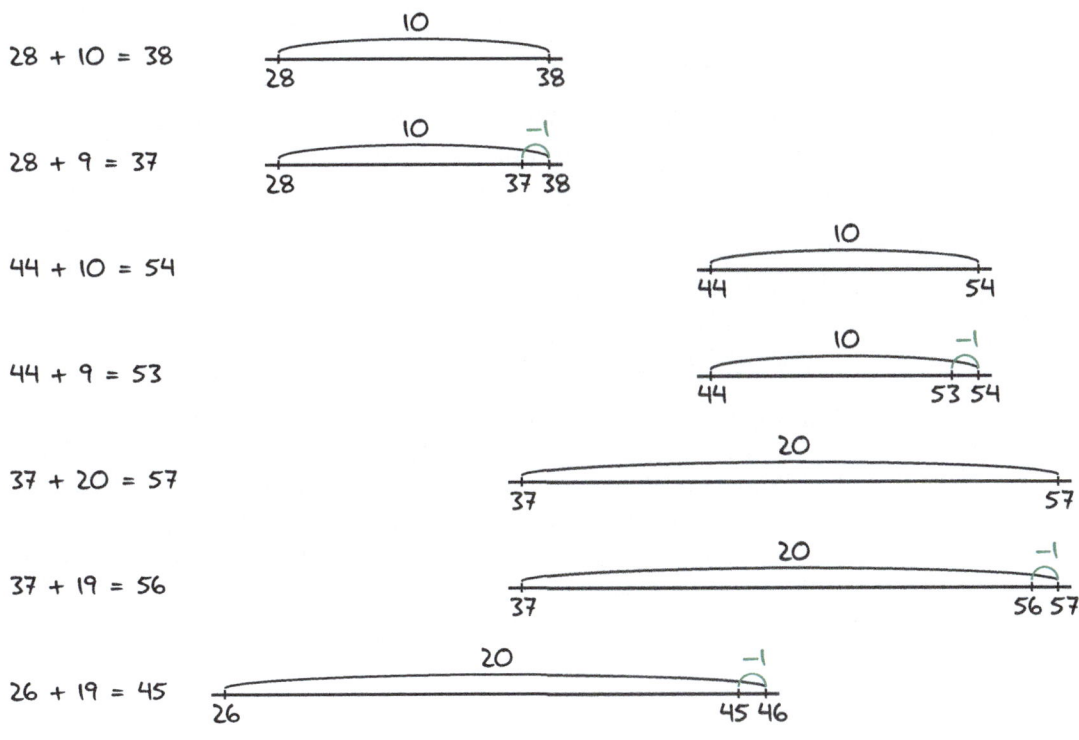

Tasks where we purposefully plan the numbers can also help students develop this strategy.

> The second grade ordered 35 cheese and 40 pepperoni pizzas. How many total pizzas did they order?
>
> Actually, they found out they only need 35 cheese and 39 pepperoni pizzas. How many is the order now?
>
> The leader ordered 44 sodas and 30 lemonades. How many total drinks?
>
> Wait, they only need 44 sodas and 29 lemonades. Now how many total drinks?

Count Arounds, as described in Chapter 8, are a great way to help students realize that they can add multiples of 10 to any number and use those results to add a little bit more or a little bit less. If you use Count Arounds while developing the Add a Friendly Number strategy, you can up the ante here. During these Count Arounds, start with a different single-digit number and add 9 each time. Line up the numbers so that students have opportunities to notice, describe, and generalize place value patterns. Discuss the place-value patterns and the places where the pattern seems to change and why.

> ### TRY IT
>
>
> Solve 56 + 28 using an Add a Friendly Number Over strategy. Represent your strategy with a number line.

IMPLICATIONS OF THE ADD A FRIENDLY NUMBER OVER STRATEGY FOR DEVELOPING MATHEMATICAL REASONING

The Add a Friendly Number Over strategy is fantastic to help students begin to think ahead. Because you are adding a number that is a bit too much, you already have a good estimate of the sum, and you know that your estimate is a bit too big. Answers are reasonable because students are reasoning using relationships they own.

Answers are reasonable because students are reasoning using relationships they own.

This Add a Friendly Number Over strategy is efficient for many bigger and more complicated numbers.

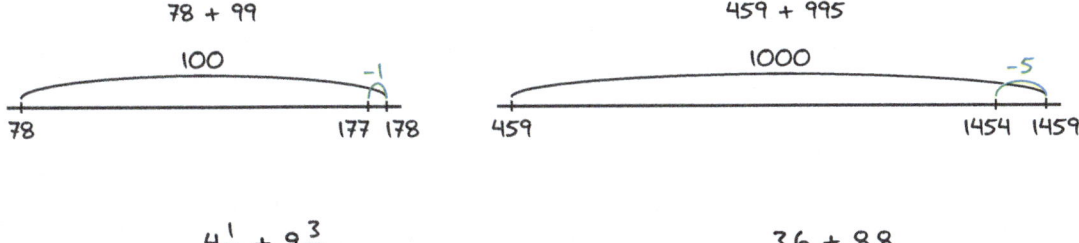

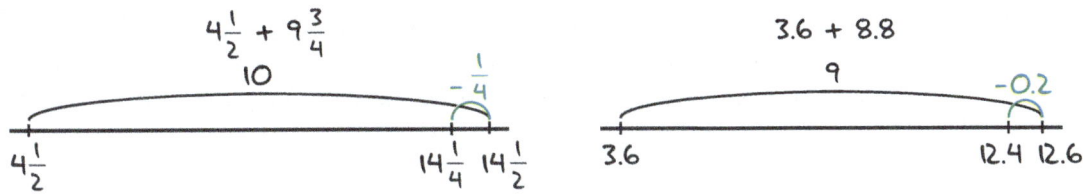

The Over strategy is also worth developing because unlike most of the strategies, the Over strategy shows up in all four operations. (See Chapter 7 for the Subtract a Friendly Number Over strategy.)

THE GIVE AND TAKE STRATEGY

Addition
Split by Place Value
Add a Friendly Number
Get to a Friendly Number
Add a Friendly Number Over
Give and Take

Source: Adapted from Math Is Figure-Out-Able at https://www.mathisfigureoutable.com/ with CC Attribution-NoDerivatives 4.0 International License

The Give and Take strategy is an equivalence strategy, which means creating a problem that is equivalent yet easier to solve.

When students consider a problem like 37 + 28, it can occur to them that 37 is close to 40. Almost simultaneously, they can wonder if it would be convenient to grab the needed 3 from the other addend, 28. Sure enough, it's easy to snag 3 from 28, leaving 25!

In context, consider adding piles of marbles. One pile of marbles that had 37 now has 40 and the pile that had 28 marbles now has 25. No one lost their marbles! And 40 plus 25 is slick, just 65. As long as we don't change the total number of marbles, just rearrange them in a clever way, the total stays the same but is easier to find.

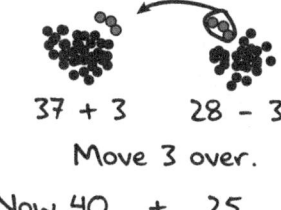

Just moving marbles doesn't change the total number of marbles.

The Give and Take strategy is the most sophisticated of the major addition strategies. This does not mean that it is necessarily the goal for students to use in all problems. Rather, it means that it requires the most brain power to give and take—it requires students to grapple with more things simultaneously and plan ahead.

The Give and Take strategy is especially nice when the given problem is tricky or seems harder to solve.

For problems where the place values roll over, like 67 + 29, giving and taking is helpful to create equivalent easier to solve problems: (67 + 3) + (29 − 3) = 70 + 26 or (67 − 1) + (29 + 1) = 66 + 30.

Eventually, students can use partners of 100 to think about problems like 67 + 89 as (67 + 33) + (89 − 33) = 100 + 56 or (67 − 11) + (89 + 11) = 56 + 100. Partners of 100 and Give and Take for the win!

Students can also use a version of Give and Take, partnered with doubles. When students work with doubles often and begin to recognize them (like 35 + 35 = 70) that can influence how they think about a problem like 38 + 34. Take 1 from 38 and give to 34, creating 37 + 35, which is 2 more than double 35, so 72.

When students become adept at Getting to a Friendly Number and Adding a Friendly Number Over, it's time to start developing the Give and Take strategy. You will probably notice some students starting to get to a friendly number and consider how much is left over at the same time. Celebrate this thinking to encourage more students to take notice.

> **TIP**
>
> Students will continue to build the Give and Take strategy in Grades 3 through 5 as they deal with larger and more complicated numbers.

Don't press this strategy too soon and never teach it as a series of steps to do. If you notice that students are randomly adding and subtracting numbers from addends without purpose or knowing how to determine appropriate, helpful numbers, it might be a sign they are trying to mimic. Remind students that the goal is use reasoning. Emphasize that students should do what makes sense to them. Math is not about doing a bunch of things to get answers. It's about sensemaking.

MODELS TO BUILD THE GIVE AND TAKE STRATEGY

Because of the simultaneity involved, do not represent Giving and Taking on an open number line. Instead use equations. An open number line can represent a sequence of jumps, whereas equations can represent more simultaneous giving and taking (Table 6.9).

TABLE 6.9 • Using Equations to Represent the Give and Take Strategy

47 + 35		
VERTICAL EQUATION MODEL	**HORIZONTAL EQUATION MODEL**	
$\begin{aligned}47 + 3 &= 50\\ +35 - 3 &= \underline{32}\\ &\;82\end{aligned}$	$\begin{aligned}47 + 35 &= (47 + 3) + (35 - 3)\\ &= 50 + 32\\ &= 82\end{aligned}$	$\begin{aligned}&47 + 35\\ &\underline{+3\;\;-3}\\ &50 + 32 = 82\end{aligned}$

HOW TO TEACH THE GIVE AND TAKE STRATEGY

Many students will not be ready in second grade to develop the Give and Take strategy. For those who are adding friendly numbers and getting to friendly numbers with few big jumps, you could engage students in Problem Strings like the one in Table 6.10.

TABLE 6.10 • Problem String Using the Give and Take Strategy

Download a handy PDF for this Problem String.

https://qrs.ly/n7gl26s

PROBLEM	TEACHER
38 + 26	"What is 38 and 26?" *Represent a couple of strategies on an open number line.*
40 + 24	"Same sum? Interesting. Anyone see any connections between the problems?" *Represent with equations.*
34 + 30	"Same sum again? What's going on? If you have two piles of marbles, what is happening with the marbles in these problems?"
45 + 17	*Represent one strategy on an open number line.*
42 + 20	"Same sum? What happened with the marbles in the 45 pile? The 17 pile? How does that help you think about the fact that they have the same total?" *Represent with equations.*
50 + 12	"What happened with the marbles this time? If you were given 45 + 17 again, might you feel like moving marbles in order to solve a friendly problem?" *Represent with equations.*
27 + 18	"Could you move marbles with this problem? How? Why? Seems pretty helpful to move some marbles to create an equivalent problem that's easier to solve."

Here's a sample final display for this Problem String:

$38 + 26 = 64$

$40 + 24 = 64 \qquad \begin{array}{r} 38 + 2 = 40 \\ +26 - 2 = +24 \\ \hline 64 \end{array} \qquad 34 + 30 = 64 \qquad \begin{array}{r} 38 - 4 = 34 \\ +26 + 4 = +30 \\ \hline 64 \end{array}$

$45 + 17 = 62$

$42 + 20 = 62 \qquad \begin{array}{r} 45 - 3 = 42 \\ +17 + 3 = +20 \\ \hline 62 \end{array} \qquad 50 + 12 = 62 \qquad \begin{array}{r} 45 + 5 = 50 \\ +17 - 5 = +12 \\ \hline 62 \end{array}$

$\begin{array}{r} 27 + 18 \\ +3 -3 \\ \hline 30 + 15 = 45 \end{array} \qquad \begin{array}{r} 27 + 18 \\ -2 +2 \\ \hline 25 + 20 = 45 \end{array}$

Because students are giving to one addend to get to a friendly 10 in double-digit addition, they are using partners of 10. Playing I Have, You Need (described in Chapter 8) with partners of 10 and partners of 100 is helpful to make those partners automatic.

Tasks where we purposefully plan the numbers can also help students develop this strategy.

> The second grade needed to order 38 cheese and 27 pepperoni pizzas. How many total pizzas will they need to order?
>
> Later that day, some students changed their minds, so they needed 40 cheese and 25 pepperoni. Now how many will they need to order? Did the total change? Why, or why not?
>
> Right before placing the order some students changed their minds again, so they got 35 cheese and 30 pepperoni. Did they get everyone's order? How do you know?
>
> They need to order 34 sodas and 19 lemonades. How many drinks?
>
> And again, some students changed their minds, so they need 40 sodas and 13 lemonades. How many total drinks do they actually need? Could you use the first order to help? How?
>
> What's another way that students might change their order that would make finding the total number of drinks easier to find?

TRY IT

Solve 56 + 28 using a Give and Take strategy. Represent your strategy with both a horizontal and vertical format. Now solve it again by making the other addend nice. Which way do you like more? Why?

FREQUENTLY ASKED QUESTIONS

Q: If the Give and Take strategy is the best, why don't we teach just it? Why do we have to teach the other addition strategies? I'll just tell students to add and subtract, and it'll be really easy.

A: Many students will be able to follow your instructions to "add and subtract" and get correct answers for a short while, but as soon as you start solving subtraction problems, they will have a new rule to memorize and mix up. Remember, it's not about mimicking steps to get answers; it's about building numerical relationships. You'll have students with those important connections *and* getting correct answers!

IMPLICATIONS OF THE GIVE AND TAKE STRATEGY FOR DEVELOPING MATHEMATICAL REASONING

Once students have developed the mental strength to consider and deal with both addends at the same time, their brains are able to consider more things simultaneously. This strategy is ideal for bigger and more complex numbers, as students consider what they can take from one addend to give to the other with decimals, fractions, and larger whole numbers.

The Give and Take Strategy with bigger and more complicated numbers

$$4.8 + 0.2 = 5$$
$$+ 3.5 - 0.2 = \underline{3.3}$$
$$8.3$$

$$9\tfrac{7}{8} + 4\tfrac{1}{4} = (9\tfrac{7}{8} + \tfrac{1}{8}) + (4\tfrac{1}{4} - \tfrac{1}{8})$$
$$= 10 + 4\tfrac{1}{8}$$
$$= 14\tfrac{1}{8}$$

$$789 + 211 = 1000$$
$$+ 465 - 211 = \underline{254}$$
$$1254$$

COMPARING THE MAJOR ADDITION STRATEGIES

Table 6.11 shows a comparison of the major addition strategies.

TABLE 6.11 • Comparing the Major Addition Strategies

STRATEGY	DEALING WITH THE ADDENDS	ONE MOVE AT A TIME OR SIMULTANEOUS	PLANNING AHEAD	WHAT IT BUILDS
Split by Place Value	Break up both addends into place value parts	One move at a time; sequential		Tens and ones place value
Add a Friendly Number	Keep one addend whole, add the multiple of 10 first, then the rest	One move at a time; sequential		The place value pattern of adding multiples of 10 at a time
Get to a Friendly Number	Keep one addend whole, get to the multiple of 10, then add the rest	One move at a time; sequential		Using partners of 10, adding multiples of 10 at a time
Add a Friendly Number Over	Keep one addend whole, add a multiple of 10 that's too big, then adjust back	One move at a time; sequential	Requires planning ahead by using a number that wasn't there to begin with	Adding multiples of 10 at a time, compensating for adding too much
Give and Take	Act on both addends by taking from one addend to give to the other	Simultaneous	Requires planning ahead, *Can I create an equivalent problem that's easier to solve?*	Simultaneity, equivalence

Each of these five strategies serves different purposes in both building Additive Reasoning and preparing learners for other more sophisticated thinking later. With robust understanding of the place value at work using these major relationships, students are primed to solve problems fluently.

> ### TRY IT
>
> Find 38 + 29 using each of the five strategies. Represent the strategy with a model.

> ### FREQUENTLY ASKED QUESTIONS
>
> **Q:** You mentioned the swapping strategy earlier. What's that?
>
> **A:** It has everything to do with why 97 + 39 = 99 + 37 (now Over strategy, anyone?). Since it's not a major strategy, I won't elaborate here, but have fun with it!
>
> **Q:** What if I have students who still revert to counting by ones, especially when the numbers get bigger? Should I just teach them the algorithm? All of this counting seems really inefficient.
>
> **A:** No, do not teach them the algorithm because they will get trapped and not progress. They need to grapple and win with larger numbers. Help bridge from their counting to larger jumps of numbers, in groups of 10. Build on their thinking.

Stephanie Lugo has just given her second-grade class word problems to work on, beginning with a combining result-unknown problem, 38 + 23. As students start working with paper and pencil, Stephanie looks at a few students' papers, encouraging students as they dig in. "Nice start," and "I can see how you're thinking."

She kneels next to Marcus, who is counting by ones on his fingers. She asks, "How are you thinking about this problem?

I am very interested to see what she will do with him. He's going to count 23 times!

Marcus responds, "I started with 38," and he holds up his fingers as he counts by one, "39, 40, 41, 42."

Stephanie quietly interrupts, "Great. I'm going to model your thinking. So, you started with the 38 books..."

|
38

Marcus picks up counting, putting up his fingers one at a time, "Yeah, and then I counted 39...."

Every time he puts up a finger and says a counting number, Stephanie echoes quietly, "So you had one more (*draws a jump and labels it with a 1*) and landed on 39 (*labels the landing spot as 39*)."

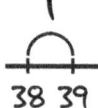

"And one more (*draws and labels the next jump of 1*) and landed on 40."

They continue this way until they've added all 23.

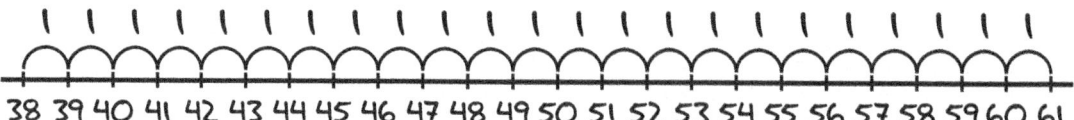

"Great job, Marcus! You've got fantastic perseverance. Whew! That took effort and you stuck with it. Nice!" With a big smile, Stephanie takes a deep breath and says, "There were so many jumps. It might be helpful to keep track. Do you mind if we keep track of some friendly chunks?"

When Marcus takes a deep breath and nods (he was working hard!), Stephanie points to the number line she's been drawing. "So, we started at 38 and made a bunch of jumps," and she counts the first 10 jumps, "1, 2, 3, 4, 5, 6, 7, 8, 9 10," and she circles the jump. "Where do 10 jumps get you?"

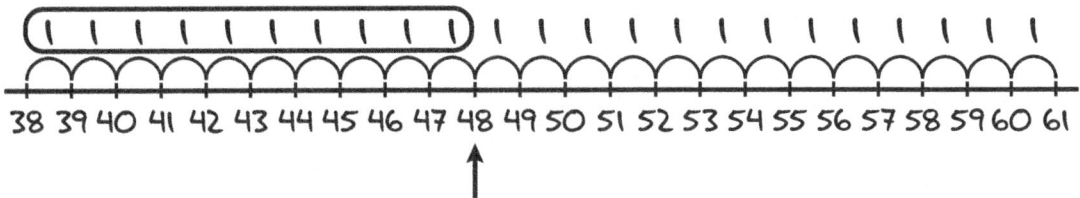

I raise my eyebrows. She's building on *his thinking*! She didn't look at his counting and say, "Stop doing that. Do this strategy instead." Rather, she is representing his thinking in such a way that then they can *build on it together.*

When Marcus answers, "48," Stephanie says, "Hum, that's interesting. Looks like 38 and 10 is 48? Cool." She is noting aloud the pattern of adding 10 to anything.

She smiles at Marcus, looks down, and begins again, "And then let's keep track of the next 10 jumps." They count together, circle the next ten, and note the ending spot: "Where did that jump of 10 land?"

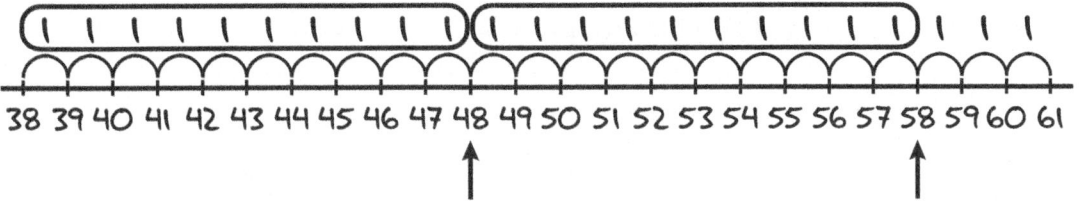

Marcus looks at the number line and looks up at Stephanie, "58. It's 8 again."

Stephaine asks, "What do you mean 8 again?"

He says, "38, 48, 58." He's noticing the pattern.

Stephanie asks, "Do you think that will happen every time we add 10? Hmmm . . . That could really be helpful. I wonder if you could use that pattern on the next problem? Want to try it?"

When he nods and digs into the next problem, Stephanie continues circulating, finding what students do know and can do and helping them develop from there.

Stephanie notes:

> Students are brilliant. They have ideas and they use them. When I can figure out what a student is thinking and represent their thinking, then we can build on it from there. Knowing the major models and strategies for addition and subtraction, the landscape with the major landmarks, has really helped me know where to nudge students next. Helping students develop as more sophisticated reasoners is just so fulfilling.

I left her classroom that day really thinking about how she was giving students credit for their effort while at the same time subtly suggesting that we could save some effort by noticing and using the patterns of adding ten to any number. She was high dosing him with that pattern and inviting him to use it.

Conclusion

You can help students develop the mathematical relationships that lead to the major strategies in this chapter, and they'll be reasoning additively with addition like champs, like math-ers! Addition, even with multi-digit numbers, is figure-out-able!

Discussion Questions

1. Which of the major multi-digit addition strategies do you already own?
2. Which of the strategies involve using partners of 10? How?
3. Enact with a thought partner: You're working with a student who is solving 48 + 29. Choose from the following prompts, and create a plausible back-and-forth focusing pattern of questioning: What are you thinking? Do you want to start with 48 or 29? Is there something friendly you could add first? Is there a friendly number nearby you could add to? How can you adjust from there?

4. Enact the following with a thought partner: You're working with a student who is solving 62 + 39. Write your own prompts, and create a plausible back-and-forth focusing pattern of questioning.

5. Solve each of the following, representing each of the major strategies on an open number line and with equations. Attempt to align the number lines and make the jumps proportional: 46 + 28, 53 + 27.

CHAPTER 7

The Major Strategies for Multi-Digit Subtraction

FIGURE 7.1 • The Second Level of Sophistication in Mathematical Reasoning

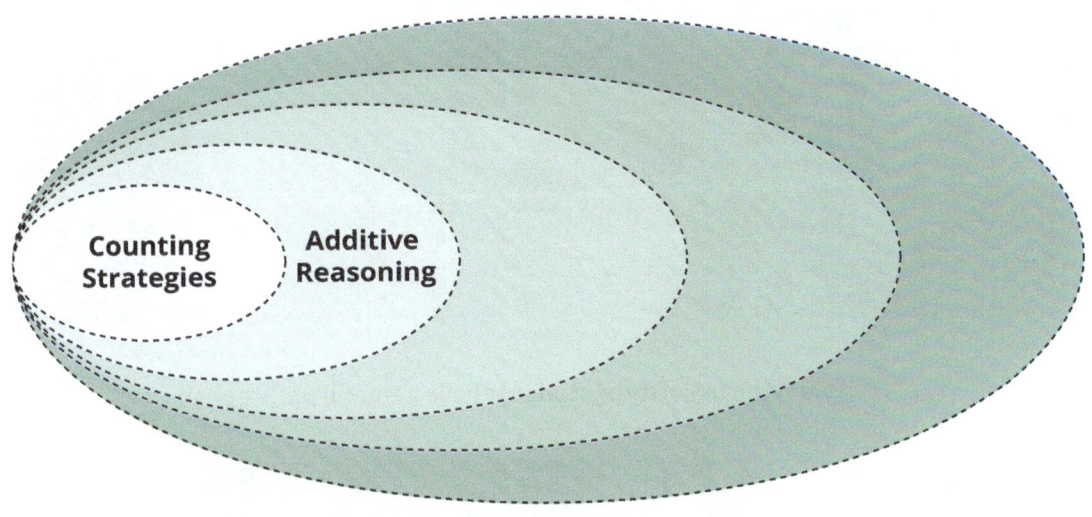

Source: Adapted from Math Is Figure-Out-Able at https://www.mathisfigureoutable.com/ with CC Attribution-NoDerivatives 4.0 International License

Multidigit subtraction brings its own challenges and opportunities. We're starting with Stephanie Lugo as she facilitates a Problem String for her second-grade class.

She begins, "Okay, here's the first thing that I want you to think about. What is 73 minus 20? (*writes 73-20 on board*) Can you signal me when you've thought about that? 73 minus 20."

When the majority of students have a thumb up, Stephanie calls, "Ava?"

Ava responds matter-of-factly, "73 minus 20 equals 53."

With a neutral expression, Stephanie asks, "Does anybody else think it is 53?"

When students nod and put their thumbs up, Stephanie continues: "A lot of people agree. Okay, 73 minus 20 is 53. Let's see. (*draws number line*) How would I represent that? What would I start on? 73? And how did you know it was going to be 53, Ava? Can you talk me through that real fast?"

Ava says, "No, I really just automatically know it. Like my body asked my brain, like, what the answers are."

Stephanie nods and says, "You just knew that pretty automatically? Okay, so if I took away 10 (*jumps 10 back on the number line*), I would be on 63, right? And if I took away 10 more, I would be on 53 (*jumps another 10 back on the number line*). So, if I take away two 10s, (*circles the two jumps back of 10*) that's the same as 20, right? Who was thinking that? Okay, so I could do two jumps of 10 and get to 53, and that would be like taking away 20."

By modeling two jumps of 10 to get a jump of 20, Stephanie helps all students have access to this first problem.

Stephanie continues with the next problem, pausing to give students time to think in between each sentence, "Alright, here's my next one. What is 73 minus 19? (*writes 73 – 19 = _____*) If we know 73 minus 20 is 53 (*writes –20 in the middle of the 2 jumps of 10 on the top number line*), what's 73 minus 19? Can you signal me when you've thought about it?"

Caleb says quietly, "64."

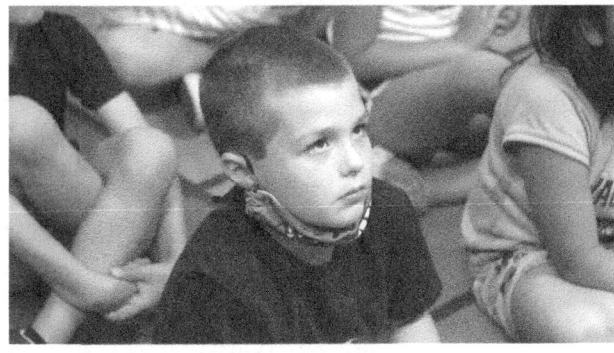

When Stephanie requests, "Say it again?" Caleb says, "I meant, uh...62."

With a neutral expression, Stephanie says, "Okay, you said 64, and then you said 62. (*Caleb nods*) Okay, so let me get you to help me here (*draws number line*). If I'm on 73, Caleb, and I take away 20, what did we just do a minute ago? We said 73, takeaway 20, was how much?"

Chapter 7 • The Major Strategies for Multi-Digit Subtraction

When Caleb and other students respond, "53," Stephanie smiles and says, "53, okay. And so, you just told me 73 takeaway 19 is 64 or 62."

Caleb very quietly suggests 54, 52, then 72.

Stephanie says patiently, "You're saying it's 72 now? Okay, do you all agree?" *many students give a thumbs-down or shake their heads.*

With a very kind and matter-of-fact demeanor, Stephanie makes it a normal part of math time to change your mind and to have classmates disagree with you. "All right, so Caleb, what do you want to do? A lot of the class isn't agreeing with you. Do you want to think through it with me, or do you want to get help and have some more time to think?"

This question, an important teacher move, gives Caleb the choice to think through the problem with Stephanie and the class or ask for help. He gets to choose.

Caleb says, "Get help."

Stephanie honors that choice. Notice that she doesn't just tell Caleb he is wrong and provide the correct answer. Instead, she is creating an environment where students can grapple with these essential relationships that build toward Additive Reasoning. Caleb's answer is incorrect, but he is still thinking. Stephanie then gets the input of other students, and there is some confusion about what the correct answer is, but she maintains a neutral expression and keeps pressing the students to clarify their thinking. As they do, the connection between the first problem (the "helper" problem) and the current problem (the more difficult, "clunker" problem) becomes more clear to the students. After a few minutes of grappling with this problem, taking some paths that are soon recognized as incorrect, Bobby suggests, "I actually have a different idea."

Stephanie responds, "Okay," and Bobby continues, "I actually think it's actually 54."

As Stephanie brilliantly responds with a neutral expression, "Okay, now you think it's 54," Bobby motions jumping on a number line and says, "Minus 1 from 20, it's 19."

Stephanie says, "Oh, okay, so you're saying we don't want to take away the whole jump of 20. (*using both hands to span from 53 to 73*) You said if you minus 1 from 20, it's 19 (*moves hands to span just a little less*). That's all we want to take away.

Bobby says, "Yeah, so add 1."

Stephanie says, "Okay. So, right here you said add 1 back. (*draws red jump of 1 to the right*) And what did you say it was?

When Bobby says, "54," Stephanie responds, "54," and writes 54 in red as the landing spot.

Stephanie continues: "So you're saying we don't want to take away the full 20. We only want to take away 19. And he said 19 is just 1 less than 20. So, that would just be this amount, right?" (*holds her hands up to the span of 19*).

Stephanie continues: "Okay, this is the 19. He got that by taking away 20, and then adding 1. Because he didn't want to take away 20, just a little bit less than 20, 19."

Stephanie asks, "Does that make sense to anybody else?"

Stephanie then asks the next two problems, one at a time. The students use the helper, 82 – 40, to reason about the clunker 82 – 39. Stephanie continues to reiterate and model student strategies, asking clarifying questions and making the thinking visible.

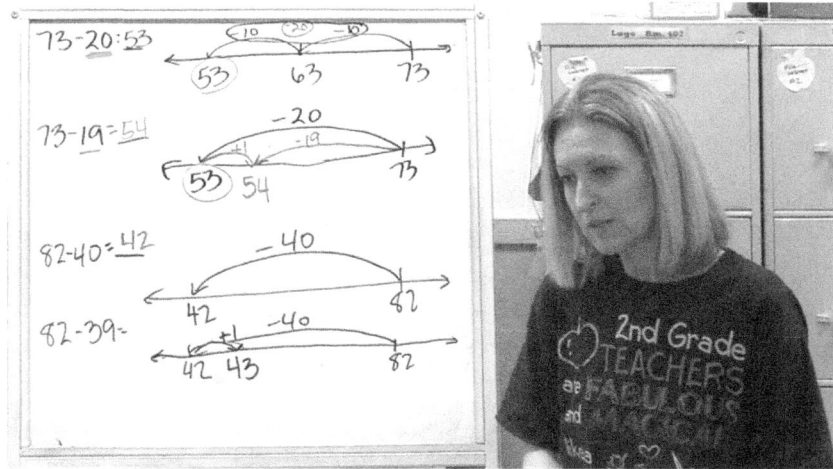

Stephanie asks the final question of the string, without a helper, "Okay. Now, I want you to think about what would 63 minus 28 be?"

Harper looks like she's concentrating, and then she smiles and says, "It's a different one."

Stephanie smiles and, nodding, says, "It's different, isn't it? But even though it's different, could you use the strategy? Could you use that going a little bit farther back (*motions to the far end of the whiteboard*), and then adjusting (*motions back toward the near end*)? Could you use that to help you? Hmm."

170 • Part III • Developing Additive Reasoning

Before you read on, try to solve 63 − 28 using the same strategy that Harper used. What relationships are you thinking about and using?

SJ says, "I know it, but at the same time I feel like I don't."

Stephanie nods. "SJ, you said, 'I think I know it, but at the same time I think I don't.' What number is . . . Could we make the 28 into something a little bit friendlier and easier to subtract? (*Stephanie's face shows intense interest*) What would you want to subtract instead of 28, SJ?"

SJ says, "I want to break it up into like 8s" (*gestures indicating several jumps*).

Stephanie respectfully acknowledges his thinking then helps the class focus on the strategy at hand, "You could. If we were kind of thinking this go a little bit farther (*motions to the far side of the whiteboard*), and then adjust back up (*motions back toward the near side*), what number is very close to 28? What are you thinking it could be, Malachi?"

Malachi says, "38."

While 38 is related to 28, it's not quite the relationship she's going for. Stephanie suggests, "Or 30. How far is 30 from 28?"

Students call out, "2."

Stephanie nods. "2 away. Would it be easier to subtract 30?"

Chapter 7 • The Major Strategies for Multi-Digit Subtraction   171

A student says, "Oh yeah!"

Stephanie says, "Okay. Do you want to try that? Could we do that? Could we subtract 30, Ethan and Lilyana?"

They reply, "Yeah."

Stephanie says, "And then, maybe move back up a little bit? Would that work?"

When Malachi nods, Stephanie continues. "If we were to do that. SJ said 28 is really close to 30. And y'all told me, 'Hey, we can subtract 30.' What would that be? What would 63 minus 30 be? Ava?"

Ava says, "63 minus 30 would be . . . (*pauses*) I forgot my answer."

Stephanie says reassuringly, "Oh, it's okay. You can figure it out again. (*Math is figure-out-able!*) You keep thinking. Let me see what Emeri was thinking. What were you thinking it was, Emeri?"

Emeri says confidently, "I was thinking it would be 33."

Stephanie says, "So, you think 63, if we took away 30, it would be 33. Okay, could we use that to help us? Okay, why don't we help each other get this one, okay? So, y'all are saying if . . . Let me move down here a little bit. (*kneels, drawing number line*) If we started on 63, and we take away 30, we're on 33."

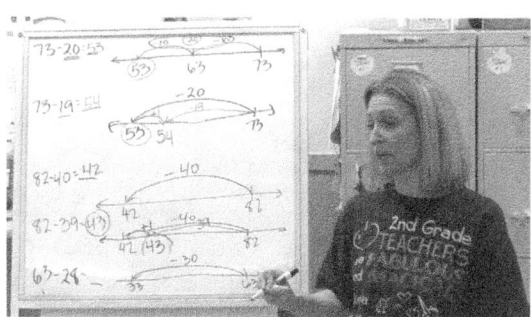

172　Part III • Developing Additive Reasoning

Stephanie says, "But we don't want to take away 30, we only want to take away 28. And you told me how close is 28 to 30?"

Students call out, "2."

Stephanie uses her hands to show the spans on the whiteboard, "Okay, it is 2 away. So, we don't want to take away the full 30 (*shows span of 30*), so, what do we do at this point? We only want to take away 28 (*shows shorter span*)."

Emeri says, "Jump 2," as she holds up 2 fingers.

Stephanie repeats, "Jump 2, Emeri says?" When students nod, Stephanie pushes for direction because this is hard for students, "Jump 2. Do you mean subtract 2 more (*points to the left of 33*) or add 2 more back (*points to the right of 33*)?"

Emeri says, "Add 2 more back."

As Stephanie draws a jump of 2 to the right of 33, she says, "Add 2 more back. And so, what would 33 plus 2 be?"

When Emeri answers, "35," Stephanie writes 35 at the landing spot as she repeats, "35? Does anybody else think that?" Lots of students agree. Stephanie ends the string putting some words to the relationships that many of the students are beginning to tinker with "Because you can turn it into an easier problem, can't you? Sometimes if you think about the numbers you're subtracting, you can think about a way that you can make it easier for you and a little bit more efficient. Okay, good thinking. I know that one took a lot of thinking. We had to help each other through it. Good job, everyone."

Reasoning in Action

https://qrs.ly/fqgl26v

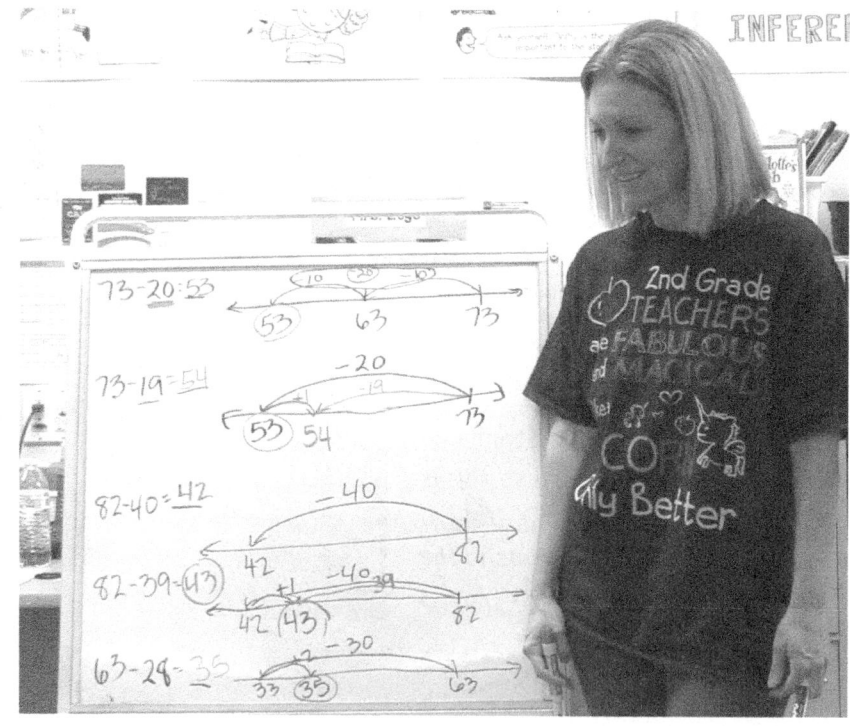

DEVELOPING MULTI-DIGIT SUBTRACTION STRATEGIES

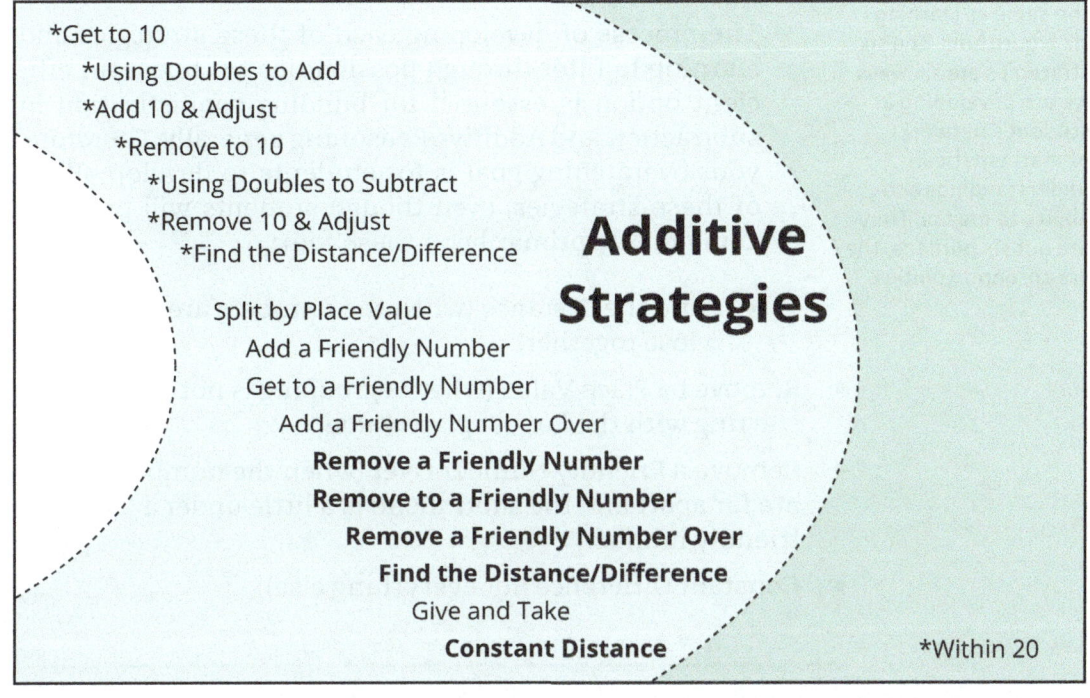

Source: Adapted from Math Is Figure-Out-Able at https://www.mathisfigureoutable.com/ with CC Attribution-NoDerivatives 4.0 International License

The multi-digit subtraction strategies develop from the single-digit strategies into more sophisticated strategies—instead of *removing to 10*, they can *remove to a friendly number* (usually a multiple of 10). Instead of *removing 10*, they are *removing a friendly number*. Also, the strategies gained in multi-digit addition will help with multi-digit subtraction. Engage students in wondering about and noticing patterns, trying to use those patterns, shifting and refining that use, and making generalizations based on their experience. This leads students to naturally and intuitively use a few important strategies.

There are six major relationships and strategies to actively help students develop for multi-digit subtraction. The following are the major strategies in Kindergarten through Grade 2:

- Remove by Place Value
- Remove a Friendly Number
- Remove to a Friendly Number

TIP

Remember, learning six strategies is not the same as learning six algorithms. The six strategies are six ways we are developing a student's network of mathematical understandings and ability to reason. They are not six burdens, they are six opportunities.

- Remove a Friendly Number Over
- Find the Distance/Difference
- Constant Difference

The process of developing each of these strategies and learning to filter through possibilities to choose an efficient option is essential for building sophistication in subtraction and Additive Reasoning generally. Therefore, your overarching goal is for students to develop all six of these strategies, even though students will probably come to rely primarily on these four:

- Find the Distance (when the numbers are close together)
- Remove by Place Value (when regrouping is not necessary), starting with the largest place value
- Remove a Friendly Number Over (when the numbers are far apart and the subtrahend is a little under a friendly number)
- Constant Difference (for everything else)

FREQUENTLY ASKED QUESTIONS

Q: What do I do with students who already use some of these subtraction strategies?

A: Some students may already be using some of these strategies—that's great! If students are comfortable with the major addition strategies and are thinking about the relationships involved, these subtraction strategies may develop naturally. Acknowledge connections students make and let them use reasoning strategies to subtract even if they haven't been formally developed as a class.

Q: Wait, aren't some of these subtraction strategies essentially the same as the addition strategies except for subtraction?

A: Basically, yes! As students build their understanding of addition relationships and build the strategies that grow naturally out of those relationships, often the subtraction strategies become intuitive extensions of their addition counterparts. Help students build on their understanding of addition by connecting these subtraction strategies to the addition strategies when possible.

> **Q:** But, Pam, my students struggle learning the one subtraction algorithm. How can I expect them to now memorize four more strategies?
>
> **A:** Remember the goal isn't memorizing and mimicking these strategies. Helping students develop the major relationships underneath these strategies and learning to reason additively with subtraction about bigger numbers is the goal.

THE REMOVE BY PLACE VALUE STRATEGY

Removing by Place Value is the subtraction analog for the addition Splitting by Place Value strategy. Just as with adding by place value, encourage students to start with the largest place value, rather than starting with the ones digit as the traditional subtraction algorithm does. Starting with the larger-value numbers helps students maintain a sense of the magnitude of the solution as they solve.

The Removing by Place Value strategy can be an efficient strategy when the subtraction of the place values stays within the place at hand, where there's no need to "borrow" or spill over into the next place value. For example, 57 – 32 is an excellent problem to Remove by Place Value because you can just subtract the place values left to right. When people do this naturally (if they have not been force-fed an algorithm), they think about the big numbers first: 50 take away 30 is 20, and 7 take away 2 is 5, so 25. This is incredibly convenient for larger numbers where the value of each place in the minuend is greater than its counterpart in the subtrahend, like 587 – 235. You can think about how 500 – 200 (300), plus the 80 – 30 (50), plus the 7 – 5 (2), is simply 352.

Think: 500 – 200 is 300 ⟶
```
  587
 -235
    3
```
Write: ⟶

Think: 80 – 30 is 50 ⟶
```
  587
 -235
   35
```
Write: ⟶

Think: 7 – 5 is 2 ⟶
```
  587
 -235
  352
```
Write: ⟶

Chapter 7 • The Major Strategies for Multi-Digit Subtraction

> **TIP**
>
> Many students will naturally start removing by place value. Don't stress it as the thing to do or students may try to do it too often, instead of developing the other major subtraction strategies.

Students will naturally gravitate toward this strategy when no regrouping is required, but encourage them to consider other strategies as well, since the relationships within the other strategies help advance students in their reasoning, both within Additive Reasoning and beyond.

Discuss with students when the strategy makes sense and what its limitations are. Encourage students to subtract left to right, considering the value of a digit rather than just thinking of it as a digit on its own. Nudge students to understand that one of the goals of working on subtraction is to increase sophistication by exploring various strategies, and Removing by Place Value has a limited use and does not provide much opportunity for rich conversation and larger connections.

THE NEXT TWO MAJOR STRATEGIES

Subtraction
Remove a Friendly Number
Remove to a Friendly Number
Remove a Friendly Number Over
Find the Distance/Difference
Constant Difference

Source: Adapted from Math Is Figure-Out-Able at https://www.mathisfigureoutable.com/ with CC Attribution-NoDerivatives 4.0 International License

> **TIP**
>
> As you listen to students explaining their subtraction strategies, you may hear students say, "I minused the 7" instead of "subtracted" or "removed." Just repeat back the more precise mathematical vocabulary as you model their thinking, "Ah, you subtracted the 7?" or "Got it, you removed the 7."

Just as with their addition counterpart strategies, the strategies Remove a Friendly Number and Remove to a Friendly Number are on par with each other in sophistication, and students need experience to develop both. You can work on both at the same time, but you may find it helpful to work on one for a while, then the other, and continue to alternate as the numbers get a bit more complex. Later, compare the two to help students get better at each.

In both of these two strategies, you keep the first number (the minuend) whole and remove pieces of the second number (the subtrahend) in sensible ways. How you break up the subtrahend is the difference between Remove to a Friendly Number (where you remove to reach the next friendly number first) and Remove a Friendly Number (where you remove the friendly part of the subtrahend first). It is all about the solver's first move. Once the first move is done, the solver might use the other strategy to remove the rest—this is another reason it is important to develop these two strategies simultaneously.

THE REMOVE A FRIENDLY NUMBER STRATEGY

Subtraction
Remove a Friendly Number
Remove to a Friendly Number
Remove a Friendly Number Over
Find the Distance/Difference
Constant Difference

Source: Adapted from Math Is Figure-Out-Able at https://www.mathisfigureoutable.com/ with CC Attribution-NoDerivatives 4.0 International License

The Remove a Friendly Number Strategy means removing a friendly part of the subtrahend and then removing the rest. As students realize they can use the patterns in our number system to think about removing friendly numbers, just like they added friendly numbers in addition, this strategy becomes natural.

For example, young students often realize that they can think about any number minus 20 (like 46 – 20) as 40 – 20 and that extra 6. If they encounter a problem like 46 – 28 they might think, "I don't know 46 – 28, but I do know 46 – 20. I'll do that and then just remove the rest. So, 46 – 20 is 26, so now I'll just remove the remaining 8." Similarly, a student might think about a problem like 64 – 37, "I know 64 – 30 so I will start with that, which is 34. Then I just need to take away the last 7."

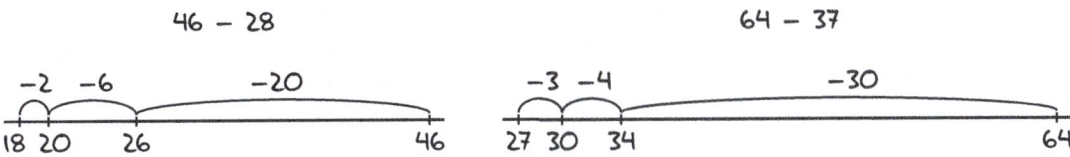

In general, the Remove a Friendly Number Strategy looks like the following:

MODELS TO BUILD THE REMOVE A FRIENDLY NUMBER STRATEGY

Represent student thinking on an open number line and, when students are gaining some comfort with the strategy, also use equations (Table 7.1). Use parentheses in equations to emphasize which part is being done first.

TABLE 7.1 • Representation of Student Thinking on an Open Number Line

56 − 27	
OPEN NUMBER LINE MODEL	EQUATION MODEL
−1 −6 −20 29 30 36 56	56 − 27 = (56 − 20) − 7 = 36 − 7 = 29

HOW TO TEACH THE REMOVE A FRIENDLY NUMBER STRATEGY

To help students both generalize and get better at removing a friendly number, engage students in Problem Strings like the one in Table 7.2.

TABLE 7.2 • Problem String Using the Remove a Friendly Number Strategy

PROBLEM	TEACHER
74 − 20	"What is 74 − 20? Did anybody think about subtracting 10 and then 10 more? We know about subtracting tens!" *Represent student thinking on a number line.*
74 − 25	"I wonder if there is anything up here that could help with this one?"
56 − 30	*Repeat. Quickly.*

PROBLEM	TEACHER
56 – 38	"Is there anything that we already know that could help us? Did anyone start by subtracting 30?"
92 – 40	*Repeat. Quickly.* "Can you think about four jumps of 10 as one big jump of 40?"
92 – 48	"Would removing a friendly number first help? How did you think about subtracting the last 8? It seems helpful to remove a friendly number and then remove the rest."

Download a handy PDF for this Problem String.

https://qrs.ly/2dgl26y

Here's a sample final display for this Problem String:

74 – 20 = 54

74 – 25 = 49

56 – 30 = 26

56 – 38 = 18

92 – 40 = 52

92 – 48 = 44

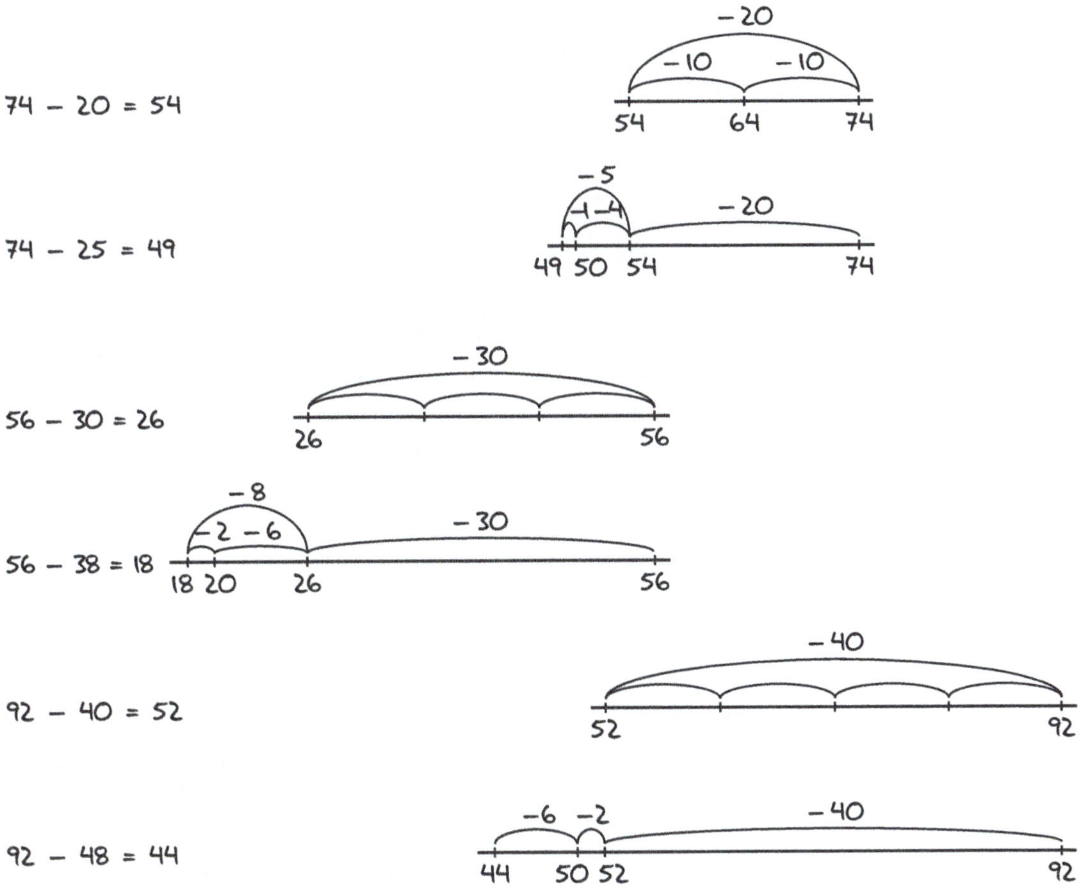

Tasks where we purposefully plan the numbers can also help students develop this strategy.

> Laila gathered 43 cool rocks from the garden. She wants to give away 20 rocks. How many rocks will she have left?

Laila actually gave away 24 rocks to her friends. How many rocks does she have left? Could you use the first problem to help? How?

The bookstore has 56 copies of a new book. They expect 30 people to buy copies during the first week. How many copies will they have left?

The bookstore sold 37 of their 56 copies. How many do they have left? Could you use their estimate to help figure it out? How?

> ### TRY IT
>
> Solve 76 – 28 using a Remove a Friendly Number strategy. Represent your strategy with both a number line and equations.

IMPLICATIONS OF THE REMOVE A FRIENDLY NUMBER STRATEGY FOR DEVELOPING MATHEMATICAL REASONING

Just as with Add a Friendly Number, Remove a Friendly Number is an important early strategy, but it becomes tedious with more digits. Notice that in the example that follows, the process involves many little jumps. After students have developed this strategy, the goal is to move on and make bigger, fewer jumps, so we'll need other strategies.

The goal is to make bigger, fewer jumps.

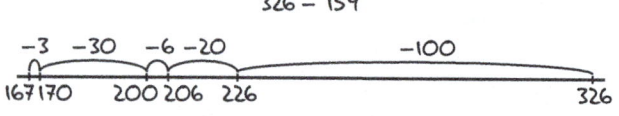

326 – 159

Start with 326, subtract 100, then subtract 59. I don't know 226 – 59 so I'll remove 20 first, then 6 to get to 200. How much is left to remove? Um, 59 – 26 = 33, so subtract 33 by removing 30, then 3 more.

Therefore, Removing a Friendly Number is less of a strategy we want added to a student's long-term toolbox and more a way to help students develop and strengthen important relationships. This strategy strengthens other essential concepts by

- helping students make sense of where numbers fit with each other: to subtract 36 and 17, keeping 36 whole, students must think about 36 as a starting point, where 36 lives, and what's around it and
- helping students generalize the notions of place value where they consider *any number minus 10* or *any number minus a multiple of 10*. Examples include 47 – 10, 84 – 10, 39 – 20, 76 – 30.

Later, students will generalize *any number minus 100, 1000*, or even decimals and fractions plus whole numbers, like the following:

- 614 – 200 to find 614 – 240
- 4297 – 2000 to find 4297 – 2009
- 3.6 – 1 to find 3.6 – 1.4
- $5\frac{3}{5} - 3$ to find $5\frac{3}{5} - 3\frac{1}{2}$.

THE REMOVE TO A FRIENDLY NUMBER STRATEGY

Subtraction
Remove a Friendly Number
Remove to a Friendly Number
Remove a Friendly Number Over
Find the Distance/Difference
Constant Difference

Source: Adapted from Math Is Figure-Out-Able at https://www.mathisfigureoutable.com/ with CC Attribution-NoDerivatives 4.0 International License

The Remove to a Friendly Number strategy means keeping the minuend whole, removing a portion of the subtrahend to get to the next friendly number, then removing the rest. Since the minuend is kept whole, students need to think about the minuend and how best to remove parts of the subtrahend to complete the subtraction.

For example, to subtract 57 – 19, a student realizes that 57 is 5 tens and 7 ones, so they start by taking away the 7 ones to get to the friendly 50. Then they determine what is left to remove (19 – 7 = 12) and remove the leftover 12 from 50, landing on 38.

TIP
When students start to become fluent with removing to a friendly number and with I Have, You Need, they can often start removing the rest in bigger chunks. For example for 45 – 28, once they have removed 5 to get to 40, they can then remove 23 in one chunk, reasoning about the relationship of 40 and 23. This is one reason to play I Have, You Need while developing this strategy.

Similarly, a student subtracting 45 − 28 can think about removing 5 first to get to 40, then removing the leftover 23 in friendly chunks.

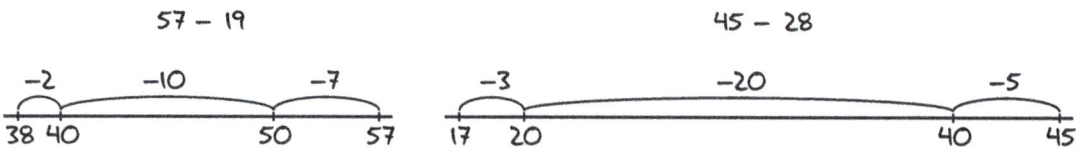

In general the Remove to a Friendly Number strategy can look like this:

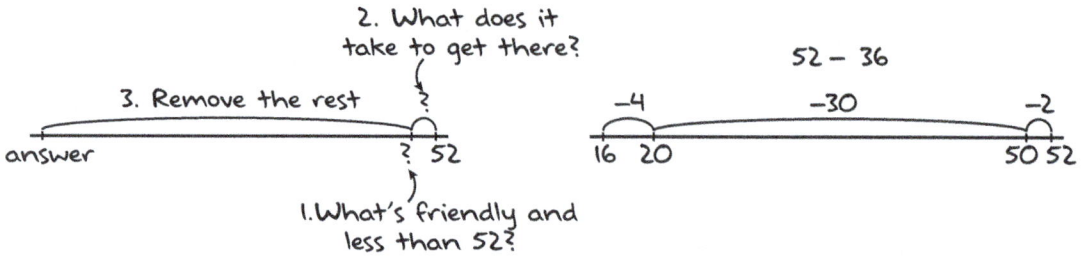

MODELS TO BUILD THE REMOVE TO A FRIENDLY NUMBER STRATEGY

Use both equations and a number line to represent student thinking. Start by representing relationships on a number line only and then with equations (Table 7.3). Use parentheses in equations to emphasize performing one part of the equation first.

TABLE 7.3 • Representation of Student Thinking Using an Open Number Line Model and an Equation Model

34 − 15	
OPEN NUMBER LINE MODEL	**EQUATION MODEL**
−11 from 19 to 30, −4 from 30 to 34	$34 - 15 = (34 - 4) - 11$ $= 30 - 11$ $= 19$

HOW TO TEACH THE REMOVE TO A FRIENDLY NUMBER STRATEGY

To help students both generalize and get better at removing to a friendly number, engage students in Problem Strings like the one in Table 7.4.

TABLE 7.4 • Problem String Using the Remove to a Friendly Number Strategy

PROBLEM	TEACHER
43 – 3	"What is 43 – 3?" *Represent student thinking on an open number line.* "40 is a nice friendly number isn't it?"
43 – 8	"What is 43 – 8? I wonder if you could use the previous problem to help you? How much more do we need to subtract?"
34 – 4	*Repeat. Quickly.*
34 – 16	"Could we use 34 – 4 to help? How?"
65 – 5	*Repeat.*
65 – 17	"Did anyone use the previous problem to help you? How? Subtracting to a friendlier number and then removing the rest seems helpful."

Download a handy PDF for this Problem String.

https://qrs.ly/2dgl26y

Here's a sample final display for this Problem String:

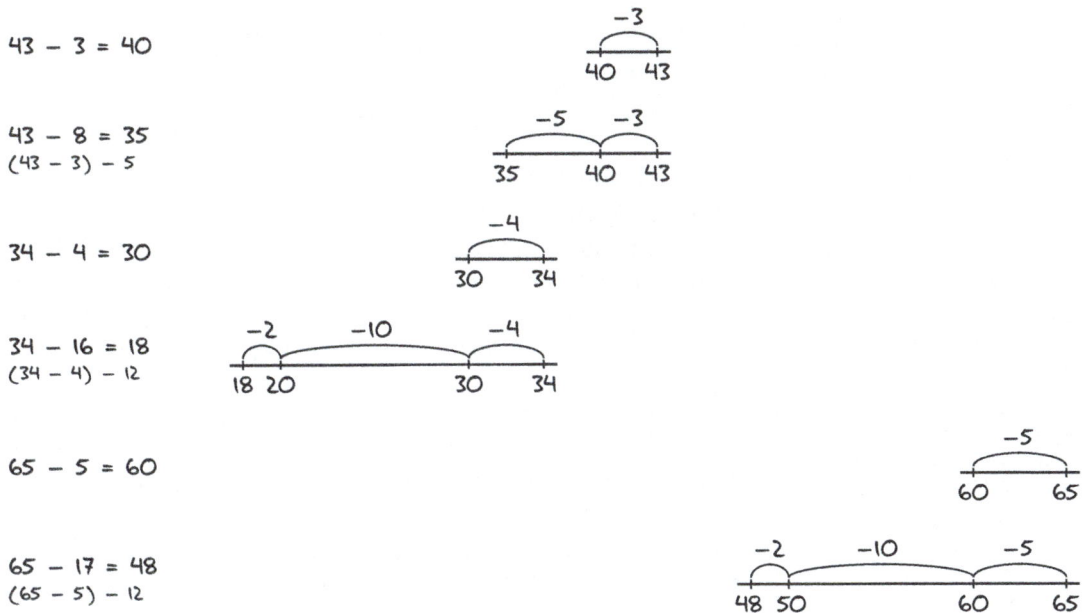

Tasks where we purposefully plan the numbers can also help students develop this strategy.

> Mira started the month with 25 pencils. She used and lost track of 5 of her pencils. How many did she have left?

TIP

Keep playing I Have, You Need with students to help strengthen those partners of 10. When you do this routine with students alongside developing subtraction strategies, it strengthens their understanding of finding the distance between two numbers as a type of subtraction. For example, "I have 65, you need . . ." with a goal of 100 is equivalent to asking "What is 100 − 65?" and emphasizes the answer as the span, or distance, between the numbers.

Mira replaced her pencils and had 25 again at the beginning of the next month. This time she used and lost track of 9 of them. How many did she have left this time? Could you use last month's total to help? How?

Elliot started first grade with 36 markers, then lost track of 6 the first week. How many did he have after the first week?

Elliot replaced the markers and got his total back up to 36 markers. Then he lost 12. How many did he have left? Could you use his first week's total to help? How?

> ### TRY IT
>
> Solve 76 − 28 using a Remove to a Friendly Number strategy. Represent your strategy with both a number line and equations.

IMPLICATIONS OF THE REMOVE TO A FRIENDLY NUMBER STRATEGY FOR DEVELOPING MATHEMATICAL REASONING

When students first start developing the Remove to a Friendly Number strategy, they tend to make many, small jumps. They think about 57 − 39 by starting at 57 and removing 7 to get to 50. Then they think about removing the rest in pieces that make sense to them, often subtracting 10 at a time and then the leftover 2.

TIP

If students are removing to a friendly number and then start to remove the rest in many little jumps, ask them, "Could you subtract a bigger friendly chunk from 50? Do you know 50 − 30?" Even if they cannot *yet*, they've heard the suggestion that it's something to try for.

You can help students begin to make fewer, bigger jumps by reminding them of their work with using friendly numbers in addition strategies. Share students' thinking who reason to remove the 39 in two or three jumps. Make the thinking visible and line up the number lines so that the smaller jumps of 10 are enveloped by the big jump of 30.

57 − 39

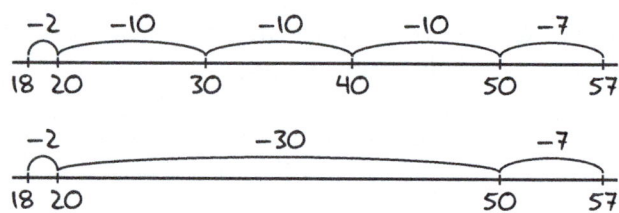

THE REMOVE A FRIENDLY NUMBER OVER STRATEGY

Subtraction
Remove a Friendly Number
Remove to a Friendly Number
Remove a Friendly Number Over
Find the Distance/Difference
Constant Difference

Source: Adapted from Math Is Figure-Out-Able at https://www.mathisfigureoutable.com/ with CC Attribution-NoDerivatives 4.0 International License

The Remove a Friendly Number Over strategy is a close relative to the Remove a Friendly Number strategy. In this strategy, students think ahead, remove a friendly number that is a bit too much, then adjust up as needed. Just like with Remove a Friendly Number, students keep the minuend whole, but when they think about removing the subtrahend, they consider friendly numbers more than the subtrahend, knowing it would be efficient to subtract something too big, because then they can add back as needed.

TIP
To help students make sense of how they should adjust after they have removed too much, you could use money. "If I owe you $17, but I give you a $20 bill, how much do you need to give me back? Why?"

For a problem like 48 – 19, think 48 – 20 = 28. We removed 1 too much, so now we need to add that 1 back in from 28 to 29, so 48 – 19 is 29.

In general, the Remove a Friendly Number Over strategy looks like this:

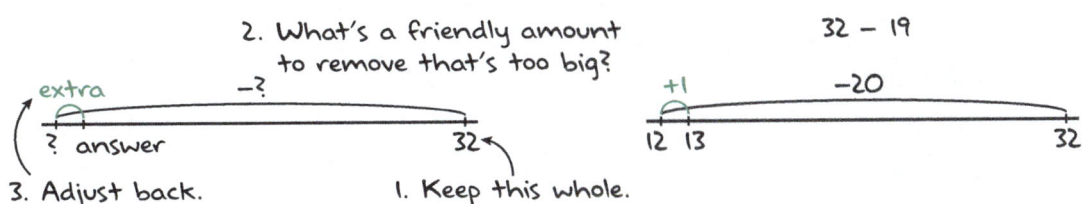

This strategy also helps strengthen the inverse nature of subtraction and addition—when you subtract a little too much, you need to "undo" that by adding back the extra you removed.

The Remove a Friendly Number Over strategy is often very efficient for problems where the subtrahend is close to a friendly number, like 39, 18, 47.

MODELS TO BUILD THE REMOVE A FRIENDLY NUMBER OVER STRATEGY

Use an open number line to represent the Remove a Friendly Number Over strategy. You can represent the jump back with a different color to show the connection between the helper problem and the final problem (Table 7.5). You may choose not to use equations for your young learners. If you do, use parentheses to emphasize the first move being made, in this case, removing the friendly number that is a little more than the subtrahend.

TABLE 7.5 • Representation of the Remove a Friendly Number Over Strategy

84 − 19	
OPEN NUMBER LINE MODEL	**EQUATION MODEL**
+1 −20 64 65 84	$84 - 19 = (84 - 20) + 1$ $= 64 + 1$ $= 65$

HOW TO TEACH THE REMOVE A FRIENDLY NUMBER OVER STRATEGY

To help students both generalize and get better at adding a friendly number that is too big and adjusting, engage students in Problem Strings like the one in Table 7.6.

TABLE 7.6 • Problem String Using the Remove a Friendly Number Over Strategy

PROBLEM	TEACHER
27 − 10	"What is 27 − 10?" *Allow think time. Represent student thinking on an open number line.*
27 − 9	"Did anyone use 27 − 10 to help you? How? Did you subtract too much or not enough? How can we adjust for that?"
65 − 20	*Repeat.* "Did anybody remove the 20 in smaller pieces? Did anyone remove the 20 all at once?"
65 − 19	"I wonder if there is anything up here that might help you? How are you making sense of which way to adjust?"
46 − 10	*Repeat.*
46 − 8	"Did anyone use 46 − 10 to help you? How? How much over did you subtract?"
33 − 19	"Did anybody create a helper problem for this one? What would be a good problem to solve first and then adjust? It seems really helpful to subtract a little bit too much and then adjust."

Download a handy PDF for this problem string.

https://qrs.ly/kigl271

Here's a sample final display for this Problem String:

27 − 10 = 17

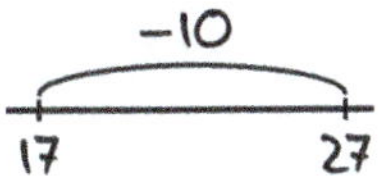

27 − 9 = 18
(27 − 10) + 1

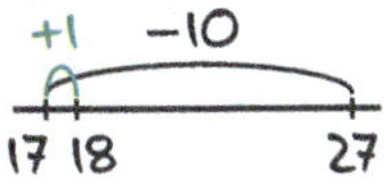

65 − 20 = 45

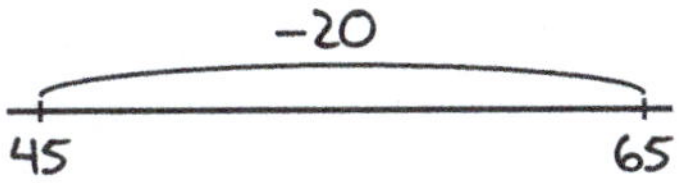

65 − 19 = 46
(65 − 20) + 1

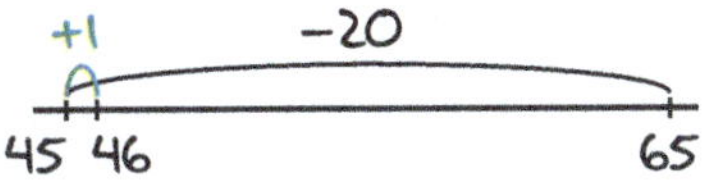

46 − 10 = 36

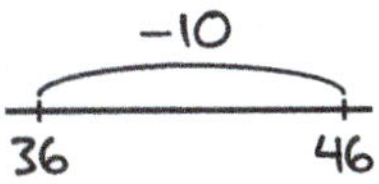

46 − 8 = 38
(46 − 10) + 2

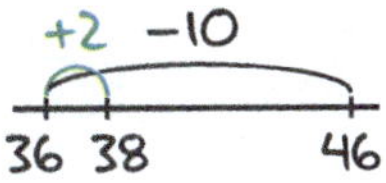

33 − 19 = 14

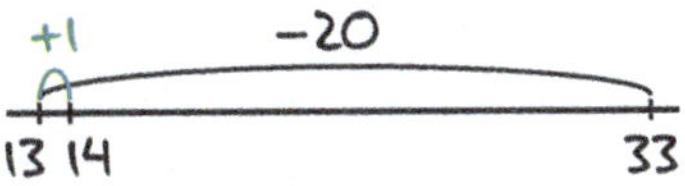

Tasks where we purposefully plan the numbers can also help students develop this strategy.

> Jorge has 35 stickers. He wants to give 20 stickers away to his friends. How many stickers will he have left?
>
> Jorge ended up giving only 19 stickers away. Since he started with 35, how many stickers does he have left? Could you use his estimate to help?
>
> Maria got 43 pieces of candy on Halloween. She planned to eat 10 of them that night. How many will she have left?
>
> Maria only ate 9 of her 43 pieces. How many does she have?

TRY IT

Solve 26 − 9 using a Remove a Friendly Number Over strategy. Represent your strategy with a number line and equations.

IMPLICATIONS OF THE REMOVE A FRIENDLY NUMBER OVER STRATEGY FOR DEVELOPING MATHEMATICAL REASONING

The Remove a Friendly Number Over strategy is fantastic to help students practice thinking ahead. Unlike a digit-focused algorithm, this strategy helps students think about each part of the subtraction problem as a value, not as a bunch of digits. Because you are subtracting a number that is a bit too much, you already have a good estimate of the result, and you know that your estimate is a bit too small. Answers are reasonable because students are reasoning using relationships they own.

The Remove a Friendly Number Over strategy is efficient for many bigger and more complicated subtrahends.

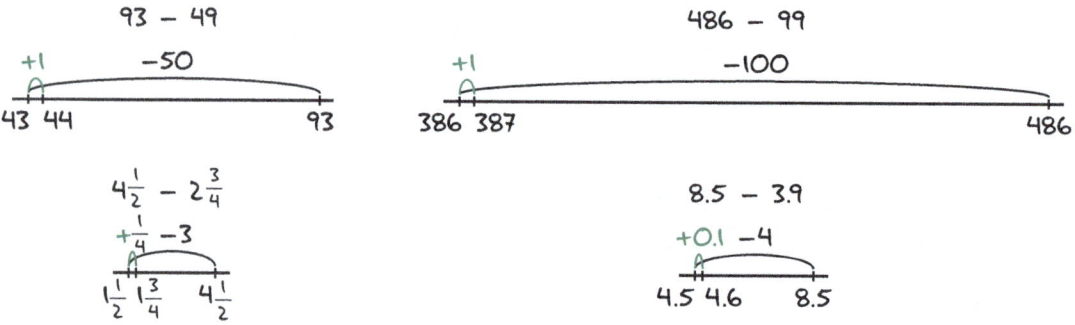

FINDING THE DISTANCE/DIFFERENCE STRATEGY

Subtraction
Remove a Friendly Number
Remove to a Friendly Number
Remove a Friendly Number Over
Find the Distance/Difference
Constant Difference

Source: Adapted from Math Is Figure-Out-Able at https://www.mathisfigureoutable.com/ with CC Attribution-NoDerivatives 4.0 International License

As we've discussed, subtraction has two meanings: difference/distance and removal. Each of the prior strategies is based on removal, where you start with the minuend and remove the subtrahend. But certain problem types beg for a distance interpretation. Similarly, certain numbers in relationship to each other beg for a distance strategy. Finding the Distance/Difference is an important strategy for subtraction.

When subtracting by finding the distance/difference between the numbers, students are thinking about the span between the two numbers. For example, the answer 36 – 29 can be found by thinking about how far apart 29 and 36 are on the number line. To find this distance, a student could either start by removing 6 to get to 30, then 1 more to get to 29 or start at 29 and add up to 36, likely stopping at the friendly 30 on the way. Either way, finding the distance is very efficient because 29 and 36 are so close together.

36 – 29 = 7

1 + 6 = 7
29 30 ─── 36

When students solve subtraction problems using the distance/difference strategy, they often use their addition strategies. Therefore, all of the information about those strategies applies here.

As students gain more familiarity with finding the distance between numbers as a subtraction strategy, they start to build an intuition about when the numbers work best for finding the difference (numbers that are close together) and when the problem works best for using a removal understanding of subtraction (when the numbers are far apart). Formalizing this understanding with students is outside the scope of this book, but is explored in the *Developing Mathematical Reasoning* 3–5 (available 2026), 6-8 (available 2026), 9–12 books (available 2027).

TRY IT

Which of these problems makes more sense to find the distance between the numbers? Which seems more efficient to just remove? Why? 83 – 7 and 62 – 59.

As students start to understand one meaning of subtraction as finding the span between the numbers, they are getting ready to explore the most sophisticated subtraction strategy, Constant Difference, which relies on interpreting subtraction as finding the distance between the numbers.

THE CONSTANT DIFFERENCE STRATEGY

Subtraction
Remove a Friendly Number
Remove to a Friendly Number
Remove a Friendly Number Over
Find the Distance/Difference
Constant Difference

Source: Adapted from Math Is Figure-Out-Able at https://www.mathisfigureoutable.com/ with CC Attribution-NoDerivatives 4.0 International License

The Constant Difference strategy is an equivalence strategy, which means creating a problem that is equivalent yet easier to solve.

When students consider a problem like 47 – 28, they might consider that subtraction as the distance between 28 and 47. Noting that 28 is conveniently close to 30, they can think about shifting that distance up 2 on the number line, to create the equivalent problem 49 – 30.

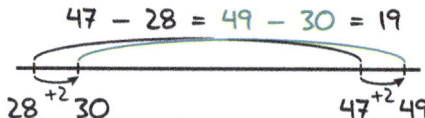

The Constant Difference strategy is the most sophisticated of the major subtraction strategies. This does not mean that it is necessarily the goal for students to use in all problems. Rather, it means that students must develop the connections surrounding the meaning of subtraction to use constant difference—it requires students to grapple with more things simultaneously and plan ahead, which most algorithms never build.

The Constant Difference strategy works well when the given problem is tricky or seems harder to solve. Constant Difference also works well when the minuend and subtrahend are not particularly close to or far apart from each other.

For problems where the place values roll over, like 57 – 38, consider the constant difference strategy to create equivalent, easier to solve problems. For example, you could think about how 38 is close to 40 and shift the problem up by adding 2 to both parts: 57 – 38 = (57 + 2) – (38 + 2) = 59 – 40 = 19. Alternatively, you could notice that 57 is close to 60 and shift both addends up 3 to get 60 – 41 = 19.

Eventually, students can use partners of 100 to think about problems like 75 – 46 as (75 + 25) – (46 + 25) = 100 – 71 = 29. Partners of 100 and Constant Difference for the win!

MODELS TO BUILD THE CONSTANT DIFFERENCE STRATEGY

Represent the Constant Difference strategy on a number line, showing both the original and the shifted problem so that students can notice the unchanging span between the two numbers. Show the jump between minuend and subtrahend for both problems, making sure that those jumps are the same size and the shift up or down is the same for both subtrahends and both minuends.

The Constant Difference strategy can also be represented using equations, but for K–2 students, the visual of the number line will likely be more effective for making the strategy visual and meaningful (Table 7.7).

TABLE 7.7 • Representation of the Constant Difference Strategy

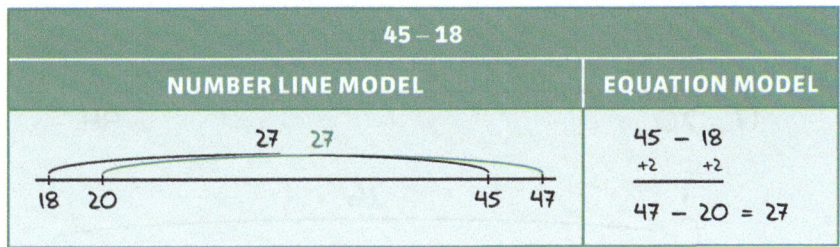

HOW TO TEACH THE CONSTANT DIFFERENCE STRATEGY

Many students may not be ready in Grades K–2 to develop the Constant Difference strategy. You could still engage students in Problem Strings like the one shown in Table 7.8, keeping the emphasis on getting better at removal or finding the distance. Model the subtraction as the distance between the numbers from each problem on the number line. Line up the number lines vertically throughout the string. When drawing the number line, make hand motions to suggest the distance relationships between the first two problems that determine where to put the tick marks.

TABLE 7.8 • Problem String Using the Constant Difference Strategy

PROBLEM	TEACHER
43 − 16	"How far apart are 16 and 43? Is that one way to think about subtraction?"
44 − 17	"Solve this problem finding the distance between the numbers please."
46 − 19	*Repeat.* "Interesting. What are you noticing? Hmm."
47 − 20	"What do you notice? 27 again? Which of these problems was the easiest to solve and why? I wonder if it could be helpful to use a friendly problem that has the same solution to think about other problems that are the same distance apart."

Download a handy PDF for this Problem String.

https://qrs.ly/elgl272

Here's a sample final display for this Problem String:

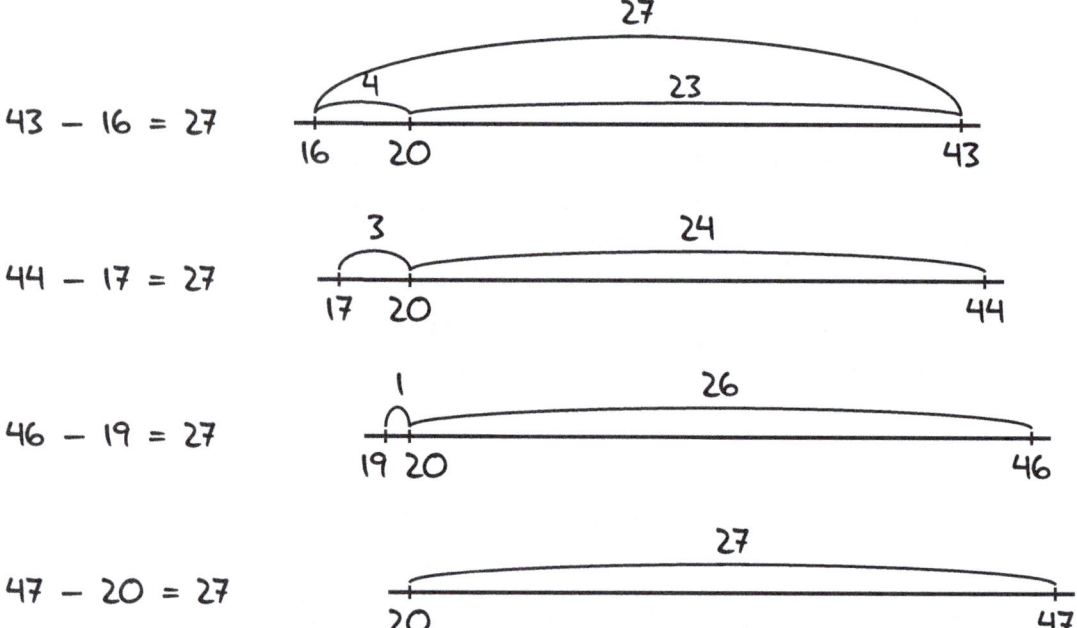

Because students are considering shifting one number to get to a friendly multiple of 10, they are using partners of 10. Playing I Have, You Need (described in Chapter 8) with partners of 10 and partners of 100 is helpful to make those partners automatic.

Tasks where we purposefully plan the numbers, using a context that emphasizes considering the span between the numbers, can also help students develop this strategy.

Paula and her friend were having a contest to see who could fold more origami hearts in ten minutes. After five minutes, Paula had folded 17, and her friend had folded 26. How far away is Paula from tying with her friend?

After two minutes, they had both folded 3 more origami hearts so that Paula had 20 and her friend had 29. How far apart is Paula now from tying with her friend? Did she get any closer?

TIP

Students will continue to build the Constant Difference strategy in Grades 3–5 as they deal with larger and more complicated numbers.

TRY IT

Solve 56 − 37 using the Constant Difference strategy. Represent your strategy using both a number line and equation model.

FREQUENTLY ASKED QUESTIONS

Q: Can students just think about the Constant Difference strategy as give–give or take–take?

A: Maybe—it depends on the mental actions that students are making. Are students just doing? Or are they reasoning about creating an equivalent problem that is easier to solve by keeping the distance between the two numbers of the subtraction problem the same? It matters. Remember, it's about creating mental mathematical relationships, not just getting answers.

IMPLICATIONS OF THE CONSTANT DIFFERENCE STRATEGY FOR DEVELOPING MATHEMATICAL REASONING

This strategy is ideal for bigger and more complex numbers, as students consider how they can shift up or down the number line to consider a friendlier, equivalent problem to solve. The sensemaking students are building from working on this strategy for whole numbers within 100 will support them as they consider subtraction with decimals, fractions, and larger whole numbers.

The Constant Difference Strategy with More Complex Numbers

$$6.2 - 4.8$$
$$+0.2 \quad +0.2$$
$$\overline{6.4 - 5 = 1.4}$$

$$8\tfrac{1}{2} - 2\tfrac{3}{4}$$
$$+\tfrac{1}{4} \quad +\tfrac{1}{4}$$
$$\overline{8\tfrac{3}{4} - 3 = 5\tfrac{3}{4}}$$

$$842 - 364$$
$$+36 \quad +36$$
$$\overline{878 - 400 = 478}$$

COMPARING THE MAJOR STRATEGIES FOR MULTI-DIGIT SUBTRACTION

See Table 7.9 for comparison of the Major Strategies for multi-digit subtraction.

TABLE 7.9 • Comparing Multi-Digit Subtraction Strategies

STRATEGY	DEALING WITH THE NUMBERS	ONE MOVE AT A TIME OR SIMULTANEOUS	PLANNING AHEAD	WHAT IT BUILDS
Split by Place Value	Break up both numbers into place value parts	One move at a time; sequential		Tens and ones place value
Remove a Friendly Number	Keep the first number whole, remove the multiple of 10 first, then the rest	One move at a time; sequential		The place value pattern of removing multiples of 10 at a time
Remove to a Friendly Number	Keep the first number whole, remove to the multiple of 10, then remove the rest	One move at a time; sequential		Using partners of 10, removing multiples of 10 at a time
Remove a Friendly Number Over	Keep the first number whole, remove a multiple of 10 that's too big, then adjust back	One move at a time; sequential	Requires planning ahead by using a number that wasn't there to begin with	Subtracting multiples of 10 at a time, compensating for removing too much
Find the Distance/Difference	Find the difference between the subtrahend and minuend	One move at a time, sequential		There are two meanings of subtraction: removal and finding the difference
Constant Difference	Shift both numbers up or down to create an equivalent problem that is easier to solve	Simultaneous	Requires planning ahead, *Can I create an equivalent problem that's easier to solve?*	Simultaneity, equivalence, solidifies subtraction as difference/distance

Each of these six strategies serves different purposes in both building Additive Reasoning, fluency, and preparing learners for other more sophisticated thinking.

> **TRY IT**
>
> Find 38 − 29 using each of the six strategies. Represent the strategy with a model.

> **FREQUENTLY ASKED QUESTIONS**
>
> **Q:** But, Pam, of course students will do your strategies during the Problem Strings because you are giving them the helper problems. What will happen when students are in the wild?
>
> **A:** First, remember that the goal is to help students develop relationships so that the strategies become natural outcomes. Second, seek to find times throughout your teaching to remind students to use what they know. For example, if they are solving word problems and the numbers are brilliant for an Over strategy, you might say, "Sure wish we could just use a bit too much. . . ." Also, notice that many of the Problem Strings give the last problem without a helper problem. When students are ready, which means many of them are starting to use the target strategy, give them a problem without a helper. Ask them to create a helper that follows the pattern in the string. Discuss why they created that particular problem and how it was helpful. This will bring clarity both to the strategy at hand and to the notion that *mathing* means to use what you know to help you solve the problem.

TIP

Once students gain enough additive relationships built by learning these strategies, they begin to think ahead and consider moves simultaneously. They consider the effect on the other number in a subtraction problem of shifting one number up or down the number line. This simultaneity and looking ahead characterize the most sophisticated of the strategies, Constant Difference.

Early in my work with Problem Strings, I was working with a select group of cutting-edge teacher leaders in my children's district. They joined in my training voluntarily because they were interested and could see the benefits.

They would periodically help me facilitate training. In our professional development sessions, I led half the time and shared a vertical approach, helping teachers understand what came before and after their grade, and the overall philosophy of teaching real math-ing. The teacher leaders led half of the time, giving on-the-ground experience of what it looks like in their grade level.

One of these teacher leaders was Ann Roman, a fifth-grade teacher. She was not only implementing what I was teaching, but she was also improving on it and writing her own Problem Strings. Ann is a remarkable, deep thinker who went on to lead professional development at the Charles A Dana Center think tank.

As we were talking one day about how to get students to use the strategies when they were not handed helper problems to use, Ann smiled and said, "Pam, I've found a great way to encourage students to use what they know to create their *own* helper! I write the string without a last helper."

She would tell students this:

> Okay everyone, let's look back at this string and talk about any patterns you're noticing. Why are many of you choosing to solve these problems the way you are? Some of these problems are super helpful to solve the next problem, aren't they? How have they been helpful? What relationships are you using?

After students put words to the strategy, getting more clarity, Ann would smile and continue.

> So, I wrote this Problem String on my way to work today. But I didn't have time to write the last helper. We just have this last clunker problem. If you'll help me create that last helper, then I'll have the string all prepared for next time. If you were to follow the same patterns in the string, what helper problem could you create for this last problem?

Ann would give the last problem of the string, and students would work to write a helper. By doing so, students are encouraged to analyze the patterns happening and use those patterns to create their *own* helper. Ann told me the following:

> Then when students run into a problem "in the wild," they'd think, "Hmmm, what helper problem could I create to help me solve this problem?" With students in the habit of creating their own helpers, they are set

TIP
Leave off the helper problem for the last problem of the string. Ask students to create their own helper, following the pattern of the string.

for life. Faced with problems, their go-to solution is to think about what they know and how they can reason from what they know to solve their current problem.

In other words, they are using what they know to reason through the problem. They are math-ing! Math is actually figure-out-able.

Conclusion

With access to reasoning about multi-digit subtraction combined with reasoning about multi-digit addition, students have the best possible start with moving on to Additive Reasoning with bigger and more complicated numbers and Multiplicative Reasoning. Even if none of their future teachers teach real math, students so instructed will know what real math is, know that they *can* figure it out, and know that no matter what else happens, they *are* mathy people. That math is actually figure-out-able!

Discussion Questions

1. Which of the major subtraction strategies do you find yourself gravitating toward or tend to think of first? Which is less obvious to you?

2. Which of these strategies involves using partners of 10? In what ways?

3. Enact the following with a thought partner: You're working with a student who does not know where to start for 47 − 29. Choose from the following prompts, and create a plausible back and forth focused on a pattern of questioning: "How are you thinking about 47 − 29? Do any relationships stand out to you when you think about 47 and 29? Could you think of 29 as 20 and 9? Does that help you think about removing 29? Can you think about 29 as almost 30?"

4. Why might you want to find the difference instead of removing? What would be true about the numbers in the problem?

PART IV

Putting It All Together

Chapter 8: Tasks to Develop Mathematical Reasoning

Chapter 9: Modeling and Models

Chapter 10: Moving Forward

CHAPTER 8

Tasks to Develop Mathematical Reasoning

"Hi everyone," I called out as thirteen second graders shuffled in and sat around the tables on the stage. They were here for "enrichment."

I had been diving into the then-elementary National Council of Teachers of Mathematics Journal, *Teaching Children Mathematics* (1994–2019), trying to learn all I could about teaching young children mathematics.

It was one of my first tries at experimenting with the things I was reading. With literally no idea of second-grade standards or expectations, I threw out a task from the journal about "digital sums." They were meant to calculate the sum of the digits representing a number. The digital sum of 62 is $6 + 2 = 8$, or for 87, $8 + 7 = 15$ and you kept going for 15, $1 + 5 = 6$. I honestly don't remember anything else about the task except that the students seemed to have some fun and I was left wondering if that was it. Throw out some esoteric thing to do, let the students mess around with it, call it enrichment.

Shortly after, I found the Young Mathematicians at Work series by Catherine Twomey Fosnot and Maarten Dolk (2001–2002). What I read about in these books seemed different than what we'd just done. Richer, more flexible, and, importantly, more focused toward important mathematics. I then found videos of Fosnot's tasks enacted in classrooms. I will never forget watching her Measuring for the Art Show videos, feeling at first that it was similar—get students engaged, chat about it a little, move on—but then realizing the depth of planning, the breadth of mathematics, the agency for the children within a tightly focused set of tasks—all these combined to create an experience unlike any I had conceived of and for all students, not just those who "needed" enrichment.

In a word, I was awestruck. And intrigued. And on fire. This felt like the mental actions my son had been doing. This was organized, focused math-ing and deliberately developing it in students.

I've spent years since then enacting those tasks, replicating them, learning from them, and from my experiences with teachers and children, building on all of that and creating my own task sequences for K–12, continually refining and adjusting. And reimaging what math class can look like and feel like as we help students (and ourselves) develop as mathematicians.

Key to this success are certain *kinds* of tasks and *sequencing* those tasks.

SEQUENCING TASKS

In the preceding chapters, you've read about developing the major counting principles, Counting Strategies, and Additive Strategies. You've seen examples of Problem Strings and read briefly about other kinds of tasks. In this chapter, you'll learn about these tasks in more detail and how to sequence them for maximum effectiveness.

When we were taught, many math classes were set up in a linear, bit-by-bit progression. You split up the content for the year in bite size pieces, teach each bit to proficiency, then move on to the next bit. Math class was a series of lessons that got harder and harder until the year was over and students moved to the next class.

But mathematics is not a linear, step by step set of small things to learn followed by the next. Mathematics is a set of interconnected and interrelated ideas, models, strategies, properties and relationships. Students can take different paths from landmark to landmark on this landscape of learning (Fosnot & Dolk, 2001). Therefore, our goal in sequencing tasks is to provide students with experiences that mutually reinforce each other in developing the various landmarks. There is a sequence, but it is of tasks that continually offer chances for students to loop back and strengthen previous ideas while advancing everyone on the map. The tasks need to be open enough that everyone has access and everyone is challenged. That means that everyone learns and everyone grows. This is challenging! The good news is that we've come a long way in developing these kinds of tasks.

> *Mathematics is not a linear, step-by-step set of small things to learn followed by the next. Mathematics is a set of interconnected and interrelated ideas, models, strategies, properties, and relationships.*

Some have called these kinds of open access tasks "low floor, high ceiling." That is part of what I'm describing, open access *enough* so that students who need an entry point have one and students who are ready for more of a challenge have that challenge. But there is more to these excellent tasks than that. These tasks also have mathematical purpose, driving toward major concepts, models, and strategies. These tasks should also have the quality of inviting students to grapple with more than one landmark at a time, helping students create that interconnected web of mathematical insight and fluency. These are not just fluffy, contrived problems to get students "engaged" so that they'll be willing to practice. This is honest, intellectual striving, inviting curiosity and creativity, argument and justification and putting forth ideas and continually refining them.

> *These are not just fluffy, contrived problems to get students "engaged" so that they'll be willing to practice.*

Japanese educators have a tradition of lesson study: working together on one lesson, coplanning, observing the teaching in one classroom, debriefing, and then enacting that lesson in the classroom. (Fernandez & Yoshida, 2004). There are many good things to say about lesson study, but the part I want to bring forward here is the immense power in focusing on one really good, rich lesson. When teachers deeply study the content of a lesson and the pedagogical moves needed to make the lesson flow, powerful shifts can occur. We can refine our teaching as we learn from our students and each other. Imagine the teacher growth centered around this work.

I propose that there is a small set of tasks that are worth this type of study–exemplar tasks where the design is purposeful and open, has an important trajectory, and uses numbers in the cleverest ways. These tasks are intriguing, truly problematic, and make the mathematics realizable (Fosnot & Dolk, 2001a). In the study of these tasks, a shared experience occurs where everyone benefits, gaining from each other's experience and expertise.

The following tasks are examples of these high leverage tasks. Thank you to Cathy Fosnot and Maarten Dolk for creating them and allowing me to bring them to you in this book.

The sequence that follows is an excerpt from the unit Organizing and Collecting (Lui et al., 2024), a unit in the K–5 series Contexts for Learning Mathematics, published by Catherine Fosnot & Associates: New Perspectives on Learning, Vero Beach, FL. (https://newperspectivesonlearning.com). The photographs are from New Perspectives Online, the video companion to the series. The materials appear here with permission.

CONSTRUCTING PLACE VALUE BY TAKING INVENTORY

Jodi Weisbart sits on the carpet with her K–1 combination class in Manhattan, New York and introduces a new context.

"I have a new investigation for you today."

She explains that the day before yesterday one of the students asked about a large container of paper bags for puppet making.

Jodi weaves a story by showing the large container of paper bags and saying, "And Issa looked at this container, and she said . . . Do you remember what you said?"

Issa smiles at Jodi and answers, "What if we had to count all of those bags?"

Jodi nods at Issa and asks the class, "What if you had to count all that? I wonder how many there are there?! That's a lot of bags! We can make a lot of puppets! I wonder how many puppets we can make?"

Jodi takes the time to talk through her thought process, making the context more transparent and realizable to her students.

She continues: "And so, that gave me an idea. I wonder how many there are? What can we do? I thought, 'That's a good investigation,' because we have a lot of things in this classroom. And not only do we have a lot of things in this classroom, we actually have a lot of people who come in and borrow things. Like Ed (*students call the teachers by their first names*) comes in sometimes, and Emily comes in sometimes, and Sandra comes in sometimes. They borrow things, and then I'm always wondering how am I going to know if they gave them all back? Like, if they borrow our scissors, I don't really know how many we have. How will I know if Sandra's class gave them all back to us? How can we find out? So, that got me thinking. Issa gave me a really good idea. Could you help me figure out how much stuff we have in this classroom so that we could know if things are missing?"

> **TIP**
> Grounding students in the context helps all students realize what's happening, increasing access.

This is a wonderful, rich opportunity that feels important and challenging.

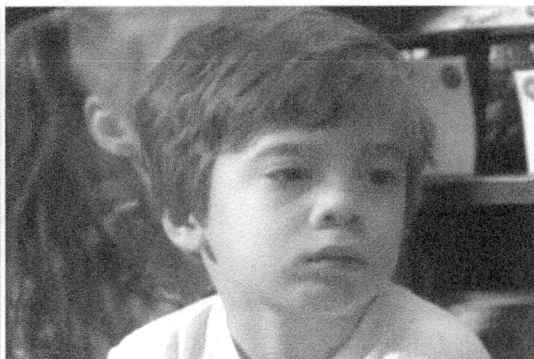

Chapter 8 • Tasks to Develop Mathematical Reasoning   207

Jodi adds some details, helping students understand the boundaries of the task and their roles: "I'm going to ask you to work with partners. I'm going to give you and your partner a clipboard and a pen, and I'm going to ask you to use your clipboard and your pen to record all the things that you're thinking about. How many things are in each book of baskets? How many blocks are in the block area? And you're going to use your paper to show me what you're thinking. And when you think you know . . . Like, Issa when you think you've figured out how many bags are in this bunch, you're going to make a sign for me. You're going to use your work and your clipboard to help you make a sign. Pretty soon, all over our room, we should have little signs that tell us how many things are in each place, and then we'll know if anything's missing."

STUDENTS WORKING

Now that the students are thoroughly involved in this important endeavor, they set out to count.

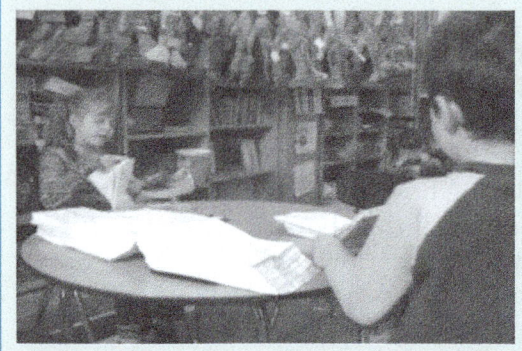
Some students count the white paper bags.

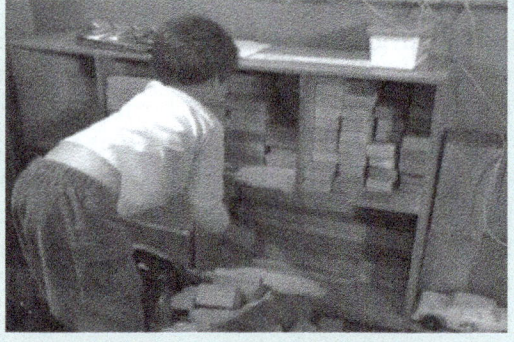
Some students count the building blocks.

Some students count books.

Some students count linking cubes.

Jodi circulates and works with students. As she listens to students count, she nudges students toward a specific strategy: keeping track of large numbers by grouping the items into packs. The groups she just *happens to suggest* are packs of 10. She provides rubber bands as the idea is raised and students begin creating packs of 10 with some loose items left over.

As the girls count 10 books, Jodie hands them a rubber band, and the girls create a pack of 10 books.

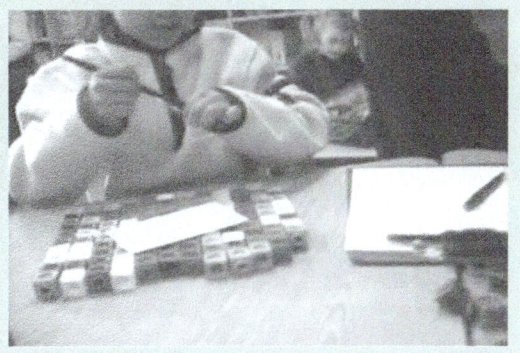

These students create snapped together trains of 10 cubes.

Students have conversations about things like what comes after 28, counting by 10s, how to write the numeral 2, what comes after 100 when counting by 10s (is it 101 or 110?), and whether to write twenty-seven as 27 or 72.

When Jodi joins Issa and Eli, they are counting paper bags by 10s as they move a pack from one pile to another.

Issa says, "170, 180, 190, 200, 201, . . . " and Eli interrupts, "210."

Both students continue counting and moving packs, and Eli announces, "240 bags."

Issa points to the 9 loose bags still on the left. "249!"

After Issa records 249 bags, Jody says, "249 bags. How many packs did you have?"

Eli says, "Um. 249."

Jodi asks, "You had 249 packs? You had 249 bags, Issa said. How many packs did you have?" Jodi puts her finger on her temple and looks curious. "How many packs of 10 do you have?"

Notice how intentionally Jodi is staying in context, in terms of bags and packs of 10 bags.

Jodi continues to confer with these young mathematicians at work (Fosnot & Dolk, 2001), artfully helping Issa and Eli count and record the total, the number of packs, and the loose bags and notice some patterns happening between the numbers.

At one point, Jodi asks, "Do you see a place in this number where the number of packs are in this number?"

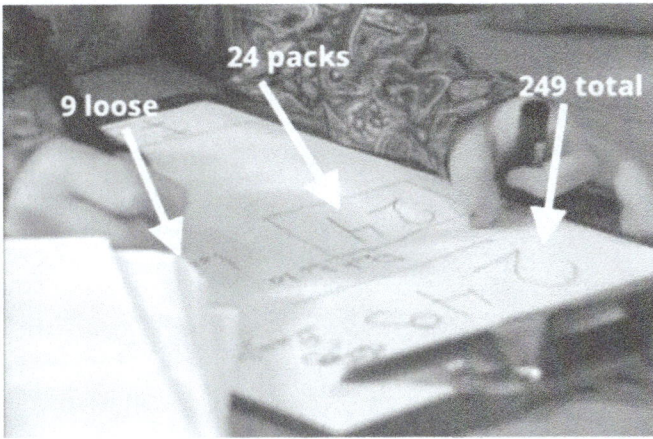

When Issa responds, "Yeah," Jodi says, "Huh. Eli, do you agree with that? That's really interesting," and then she asks for a generalization: "Do you think that if we worked with some other really big number, the same thing would happen?"

This gives Issa and Eli something to think about as they create a sign that shows how many packs, how many loose, and how many all together.

Other students are also counting and creating signs. Jodi circulates and works with groups. Many groups are counting smaller numbers of items, working on the counting sequence and beginning to consider counting by 10s.

THE NEXT DAY: A MINI-CONGRESS

The next day, Jodi gathers the students on the carpet with their clipboards for a mini Class Congress to share some of their thinking.

Jodi asks students from the block group to share, and the discussion is about grouping two 5s to make 10.

Jodi then asks Issa and Eli to share the pattern they were finding. As they talk about packs of 10 and loose and the pattern, Jodi starts a poster of "Packs, Loose, Total."

With student input, she asks questions like "If you have 100 total, how many packs do you have?" and she fills in several cases, like 100 under total and 10 under packs. Students are starting to cement that they can count by 10s with the packs of 10 they've created to find the total number of items. As a class, they have not yet brought in how the loose relate.

Over the next week, the children continue to spend time counting and recording total, packs, and loose on signs all over the classroom.

ONE WEEK LATER: A MATH CONGRESS
One week later, Jodi gathers the class for a Math Congress.

Jodi introduces the purpose of the Math Congress: "Yesterday, you were working, and working, and working on finishing your inventory, and I thought maybe we should start putting it up in one place because these signs, sometimes they fall off. If we put them up in one big place, and we keep this chart, we'll really always know how much stuff we have in this classroom. And I need you to help me."

Then Jodi gives some expectations. "When you're telling me about what you found out in your baskets, and your Band-Aids, and your blocks, make sure that you're saying it so loud and so clearly that everybody else can understand. And if you have a question about what somebody said or what they found out when they did their inventory investigation, then you make sure you stop them, and you say, 'I'm confused.' Does that make sense? Okay, so you're not only going to tell me, you're going to tell all of your friends. And your job is to make sure that you understand what they're saying. Okay?"

Jodi brings in the students' prior experience: "I noticed when I walked around that a lot of kids were doing things in packs. So, when I thought about making this list, I thought we'd have the item. That's the thing that you inventoried. The books, or the blocks, or the bags, or the Band-Aids. That's what people call it in the store. The items. The total number that you found and then how many packs you found. And sometimes how many loose ones you found. And I was thinking a lot of kids were using packs to help them keep track, so maybe we could put that up on our chart, too. That might be some information that we could have."

Now Jodi asks students to share. "I'm wondering if . . . Issa, you and Eli. Here's your clipboard. You and Eli started your investigation with paper bags, so I'm going to put 'Paper Bags.' And I'm wondering how many total you found. Can you use your clipboard to remind yourself? How many bags did you find altogether?"

Issa checks her clipboard, and Jodi fills in the appropriate cells with 249 total, 24 packs, and 9 loose and notes that the packs were "packs of 10."

They move onto other items. When Liana reports that Basket A had 40 packs, 2 loose, and 42 total, Jodi asks if Liana could show everyone those 40 packs. They get the basket out, look at the packs, and count them, and Liana decides that there were actually 6 packs.

Jodi asks the class how many books 6 packs is, students talk to their partners, and then Sequoia answers, "I think it's 60," and when asked why, she says, "I counted by tens."

Jodi goes back and forth with a few students who are not sure how many books there are until the class agrees that in the basket there are actually 61 books, 6 packs, and 1 loose, and Jodi fills in the chart.

Jodi continues, asking students to report the number of items, packs and loose, for the things they counted. They check, recount, and change some of the reported numbers, and the chart gets filled in. Students are counting packs and connecting the number of packs to the total number of items. Students are invested—at one point, Declan asks when he'll get to tell about the number of blocks he inventoried.

While they are talking about a basket of books that has 23 books, Jodi has filled in 2 packs and 3 loose, and the class is talking about how they can know how many books that is. Elijah shares: "Because I know that 10 plus 10 equals 20, and 3 more is 23."

Suddenly, Cosmo points to the chart and says, "The board!"

Jodi asks, "Put it on here?"

Cosmo excitedly says, "No, the board!"

Jodi asks, "What?

Still pointing, Cosmo exclaims, "Look! 23!" He's noticing that the 2 packs and 3 loose written on the chart correspond to the 23 total.

Jodi says, "Oh, it says it right here?! Say more."

Gabrielle calls excitedly as she points to the chart, "Yeah, 17! And 51! And 52! And 64!"

Jodi looks very interested and asks, "Are you noticing something? Talk to somebody next to you about what you're noticing."

Then Jodi asks, "So, who wants to share what they have told their friend? Sequoia, tell me what you're thinking. What are you noticing?"

Sequoia says, pointing first to the total and then to packs and loose, "There's 5 and 2 over there, and 5 and 2 over there. And then 5 and 1 over there, and 5 and 1 over there. And 1 and 7 over there, and 1 and 7 over there. And 2 and 3 over there, and 2 and 3 over there."

Rachel adds, "I noticed that every number in a row, it makes a certain number."

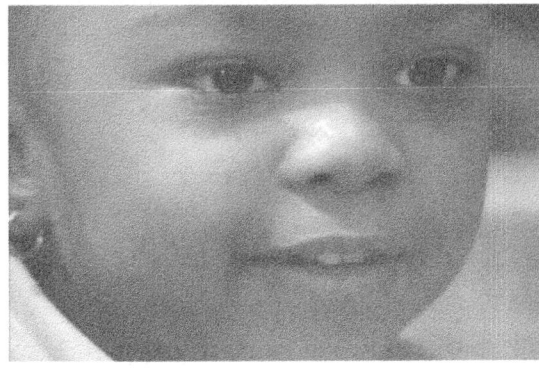

With some more back and forth, Jodi presses for a generalization, "Why do you think that's happening? Do you think that this is always going to happen?"

When Eli responds, "Not sure," Jodi notes, "Not sure, yet. What would you need to convince you? Should we find out what some more kids did and see if it happens some more?" These are brilliant questions to invite student wondering and generalizing.

The Class Congress continues with a group who counted snap cubes, and after some checking, they find that there are 14 packs, 140 snap cubes (*written by the students as 10040*) and 0 loose.

Gabrielle waves her arms and exclaims, "You did it again!"

Jodi smiles and says, "I didn't do it! It's just what happened! Do you think with packs of 10, this is always going to happen?"

A student calls out, "It is going to happen!" The excitement is palpable.

Sequoia says, "I think it's always going to happen if you even do it as you go past 99. And I think it will always happen because the numbers keep doing it."

Rachel says, "I think it's always going to happen is because every number makes sense together."

Jodi then takes a fantastic twist as she says, "In this basket, it says that there are 46 books. I'm wondering if you think, based on what you have discovered, if you think you could tell me how many packs there would be and how many loose ones in 46 books?"

BAM. Now her students get the opportunity to cement the relationships they are finding by going backward from the total number of books to the number of packs and loose.

They count and check and record in the table 42 total for 4 packs and 2 loose. Then they turn to a basket with books where the students had recorded 1 pack, 2 loose, and 14 total. When Jodi writes these numbers on the chart, a student calls out, sounding disappointed, "It didn't do it again!"

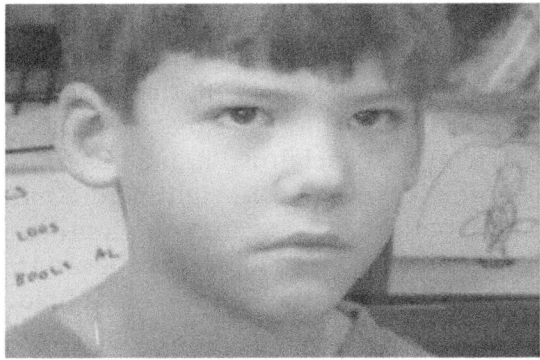 

Jodi looks puzzled and asks, "So, what do you think about that one? Do you think that that's okay? Talk to somebody next to you. Is this okay?"

Students talk in pairs. Then Jodi brings everyone back together, "Some kids are saying, 'Ah, so what? The pattern didn't hold up.' Some people said, 'It can't be like that.' Some people said, 'Eh, it's fine.' What do you think about this? Issa, I heard you start to say something. Talk about that."

Issa quietly and confidently says, "It might be a mistake (*because students counted incorrectly*). Or, like if the 14 might be a 12. Or the loose 2 might have to be 4."

When Jodi records those possibilities, a student calls out, "And then it would be right!" This sounds so hopeful—they want the pattern to hold true.

They grab the basket and check it out, counting the books and packs, and sure enough, there was a mistake, and Jodi fixes the chart to read 12 total, 1 pack and 2 loose.

The very last thing you hear on the video is Gabrielle exclaiming, "It did it again! I told you, it did it again!"

I have shown this video to hundreds of teachers, and every time they lean forward with rapt attention when we hit the midpoint where Cosmo says, "The board!" and then later teachers cheer when Gabrielle waves her arms and exclaims, "It did it again!"! We are so enamored to watch learning occur right before our eyes. Pattern *finding* and then pattern *using* as students continue to grapple going from the total to packs and loose and then with the mistaken sign that made it look like the pattern didn't hold. We hear the joy as Gabrielle exclaims at the end, "It did it again! I told you, it did it again!"

Jodi intentionally designed a sequence of tasks to further the learning and differentiate for her students:

- Jodi naturally differentiated by choosing which items student pairs were to count: smaller sets like large blocks or a basket of books for students who needed more support and larger sets like small snap cubes or paper bags for students who needed more challenge.

- She gave students items such that the amounts were challenging to keep track of.
- She had rubber bands available.
- She interacted with students as they worked, complimented them on their persistence, and nudged them to create packs of 10.
- As students were creating packs of 10, Jodi suggested that it would be helpful to keep track of the total number they counted, the number of packs and the number of loose items. Even though this was planned from the beginning, she waits until it makes sense to add this instruction
- She created two Class Congresses, where students shared their reasoning about specific questions.
- Jodi asks generalizing questions like "If we know this is the number of packs and loose, do we know the total? If we know the total, can we find the number of packs and loose? What is the pattern you're seeing? Do you think this will always happen?"

Notice that this task is not random exploring. Jodi does not tell them to go "figure out place value" and hope they come back having *discovered* it. Students are clear on their task, they know what to do, and in that security they are freed up to notice patterns, make conjectures, get better at counting, and learn important things about place value with the intentional guidance from their teacher.

CONSTRUCTING THE OPEN NUMBER LINE BY MEASURING FOR THE ART SHOW

The sequence that follows is an excerpt from the unit Measuring for the Art Show (2024), a unit in the K–5 series Contexts for Learning Mathematics, published by Catherine Fosnot & Associates: New Perspectives on Learning, Vero Beach, FL. (https://newperspectivesonlearning.com). The photographs are from New Perspectives Online, the video companion to the series. The materials appear here with permission.

Let's join Hildy Martin, a teacher in New Rochelle, New York, and her second-grade class. The sequence is a Rich Task, Math Congress, Follow-up Task, and a Problem String.

RICH TASK: MEASURING FOR THE ART SHOW INVESTIGATION

The teacher, Hildy Martin, gathers the children on the carpet to introduce the task.

She announces they'll be having an art show at the end of the year, creating art on all different sizes of papers they have in the classroom, like the very large yellow paper, the smaller purple paper, and so on. Her father's friend is willing to cut paper strips to label each piece of art, but they need to create the plan for him.

"You're going to do the measuring," she said, "because I'm too busy."

The children are grouped into pairs, given a pile of two colors of cubes to work with, and go off to do the measuring.

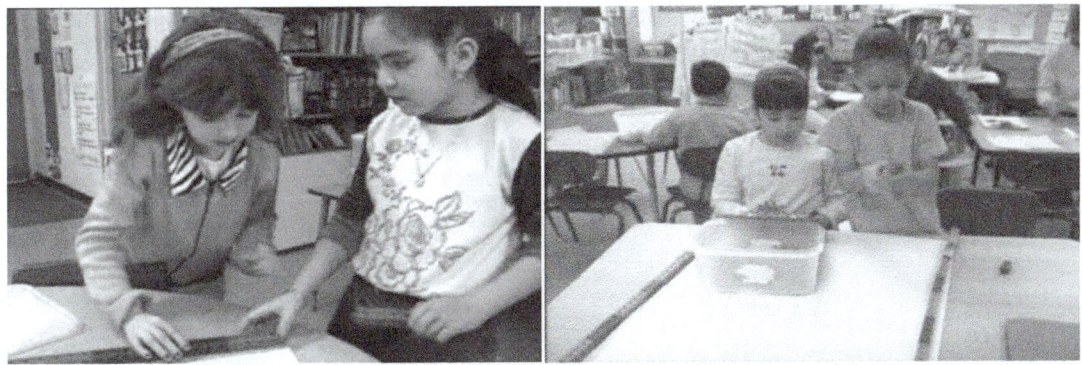

At first glance, this task seems to be about measurement and counting. And it is. But this is the start of a series of tasks. Watch where it goes.

MATH CONGRESS: MEASURING FOR THE ART SHOW

After children have measured the papers with the cubes, Hildy gathers the class again, this time to create "the plan" on a narrow strip of paper. This can be called a Math Congress, where students come together and bring their work to their classmates to create shared meaning. She asks students for the length of the short side of the blue paper. The children consult their clipboards where they noted their measurements.

Hildy explains that she's going to use the cubes to help her know where to mark as she gestures to the strip on the board. The strip that just happens to conveniently have a train of alternating groups of 5 cubes hanging above it. Wait. That cube train was hanging there in the beginning. Interesting! In fact, there were several preplanned things that nudged children to measure the papers with alternating groups of cubes. And now they'll use that structure to help them locate measurements on the cube train and mark the corresponding places on the paper strip.

"Angelica, if I'm going to use the cubes to help me, where should I put 10? Can you help me?"

Angelica says, "In the green and the white."

With Angelica's guidance, Hildy uses the pointer and chalk to draw on the board and the marker to mark the paper strip with the number 10. Hildy says, "I'm going to write it, so he'll know blue is 10, right there." She points to the mark on the strip and to the edge of the first 10 cubes in the cube train.

This continues with Hildy asking about the long side of the purple paper (22), "Gee, where should I mark 22?"

The children go back and forth with Hildy, explaining three different ways of using the groups of 5 cubes to find a length of 22, like Emily: "I know that every block, every white and green are blocks of 5, if you count 5, 10, 15, 20, and then you just count on 2 more, 1, 2."

Chapter 8 • Tasks to Develop Mathematical Reasoning   225

Then Hildy repeats this back and forth with students for long side of the cardboard (35).

In particular, Josue says, "Three jumps of 10, put the 5 and 5 together, that's 20, then 5 and 5, that's 30," and Hildy marks 30. "Just 5 more, like 10 take away 5. You have 10, that's 5 and a 5. You're just taking a 5 off." And Hildy marks 35.

Things shift when students report three different measurements for the chart paper: 45, 44, 43.

Hildy says, "Why don't we check it out?" and holds up the chart paper, reinforcing measuring principles by making sure the paper lines up properly. But this is also an opportunity for more students to consider how the alternating groups of 5 cubes can help, both to measure and to find numbers on the number line.

They decide the length is actually 46.

At this point, most students are using the alternating groups of 5 green and 5 white cubes to count by 10s to find measurements. As students find higher measurements (like 66, 84), they sometimes start from the beginning. When they do, Hildy wonders aloud if they could start from a place they know.

These children are co-constructing the number line. It is about a continuous measurement model, but it is so much more than measuring. They are dealing with span and length, nearness and neighborhood, and friendly numbers and place value.

FOLLOW-UP TASK: MEASURING FOR THE ART SHOW

The next day, Hildy begins with, "After our work yesterday, I went home last night, and I shared the work that we had done with my father's friend who is getting right on the job of cutting up those strips. But I discovered that I had some other papers at home. And I started to measure them all out with cubes, also. And one of the papers that I had at home was 19 cubes. And I had one that looked almost like it, but I wasn't sure so I took the 19, and I put it on top of this other paper and it turns out that it was 1 more."

"So, do I need to measure it?"

The class responds that no, it just makes 20, and Hildy finishes the equation.

Hildy continues: "I started to think. We have lots of papers. If we were putting them together for the art show and there was one right next to the other, I was kind of curious about how those 2 strips would be together. How long on the bottom would those 2 papers be together?"

In this moment, watching this video, I had an epiphany. Not only had the students constructed the open number line, but now they were also going to use that context to reason about addition! Oh my! I couldn't wait to see what the children did.

> *Not only had the students constructed the open number line, but now they were also going to use that context to reason about addition!*

Hildy writes 19 + 21 for the two papers and asks, "I was kind of wondering, how big the strip would have to be to go under both of those together?"

After think time, Hildy calls on students who haltingly explain the following three strategies. As they explain, Hildy draws out the relationships by slowing them down, asking questions, moving cubes, drawing jumps on the strip number line, and writing equations.

Angelica split the 21 into 20 and 1 and the 19 into 10 and 9. She added the $10 + 20 = 30$ and then added the $9 + 1 = 10$, and added the 30 and 10 to get 40.

She was Splitting by Place Value.

Amira started at 21, added 10 from the 19 to get 31, then the rest of the 9 to get 40.

Amira was Adding a Friendly Number.

 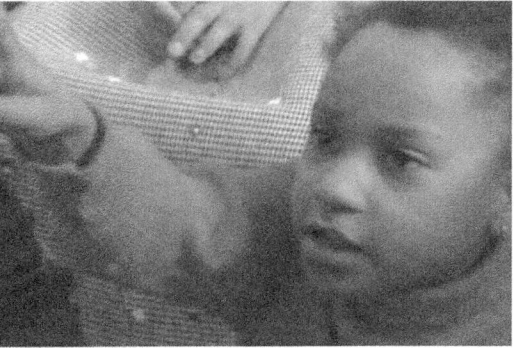

Emily kept the 19 whole and added 1 from the 21; that got her to 20. Then she added the 20 and 20 to get 40.

Emily was Giving and Taking.

These students are using the relationships to solve the problems! Notice they used three of the five major addition strategies. Amazing!

PROBLEM STRING: DOUBLE-DIGIT ADDITION

Now, if all of that isn't cool enough, we then get a glimpse into Hildy's class a few days later, where Hildy does a Problem String.

She asks students to solve each of these problems, one at a time. Students solve the problem and share their thinking, and Hildy represents their thinking on the board.

$$43 + 20$$
$$63 + 30$$
$$62 + 39$$

For each of the problems, students work with keeping the first number whole, adding friendly numbers, decomposing and composing numbers in friendly ways, and talk about their thinking, not about steps. Hildy represents their thinking on the board, making it visible, discussable, and comparable. For the third problem, 62 + 39, Hildy asks five students to share.

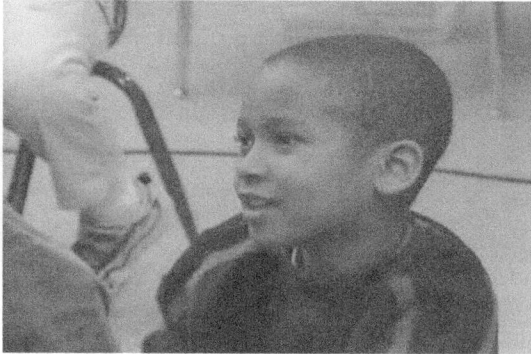

 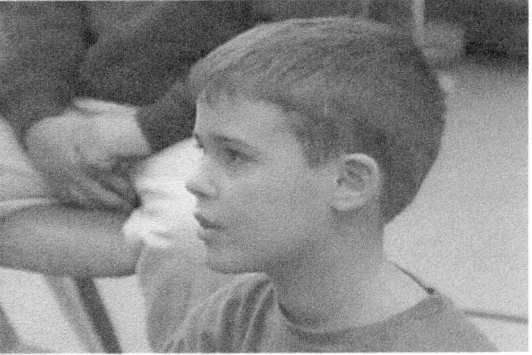

One of them is Emily, who "kept the 39 whole . . . and then I took the one (from the 62) and made a 40. . . . It got me to a friendly number." Emily is solving 39 + 42 by thinking about the equivalent problem 40 + 41.

Hildy finishes with the last problem. "Okay, I'd like everyone to think about 54 plus 48. Maybe you could think about Emily's cool way and if that could help you do 54 plus 48. And when you're ready, please put up a thumb. I wonder if Emily's cool way could help us? I see people really thinking about it."

Janet: "I knew there was 50 in that 54. And in the 48 there was a 40, and I put it together and that was 90. And then I took away from the 4, 2 and put it with the 8. And the 90 and that 10 is 100. There's 2 left. 102."

(Continued)

(Continued)

Sarah: "I know that 4 and 8. I took the 4 from the 54 and the 8 from the 48. And I know that 4 + 8 is 12. Um, and so I took the 10 from the 12 and put it, first actually I put the 50 and the 40 together. That equaled 90. I put the 10 from the 12 on the 90 and that is 100 and then I put the 2 on, so it's 102."

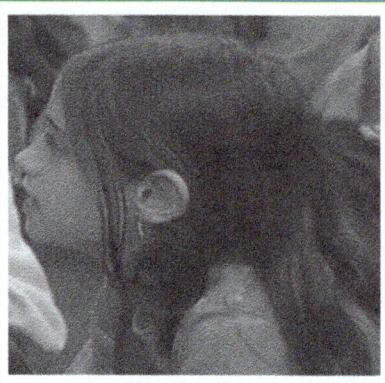

Lucero: "I held 48 in my head. And I took 2 away from the 54. I took the 2 and I got up to 50 cuz it's a friendly number. And then I took a jump of 50, and I got to 100. Then I only had 2 so I only jumped a 2 so I got to 102."

Lucero has picked up on Emily's strategy, getting that 48 to the friendly 50 and then adding the leftovers.

In this 20-minute Problem String, we see students reasoning additively, using what they know and learning in the process. We see Hildy drawing out relationships and making them visible on the board. We see her asking questions in key moments, "Where did you land when you jumped that 10?" "How many do you have left?" We see purposeful teaching and we see *math*-ing.

How did Hildy achieve this *math*-ing? Excellent teaching, a vision of what *math*-ing is, and a purposeful sequence of tasks:

- Investigation/Inquiry cube-measurement task
- Class Discussion/Congress to construct the number line
- Related Problem Talk of putting papers together
- Problem String of addition problems

This sequence helped students construct the mental map of the number line and use that mental map to reason about problems, first in the context of lengths of papers and then in the Problem String of context-less addition problems.

Fosnot (2024) has since refined and added to this sequence in the Measuring for the Art Show: Addition on the Open Number Line.

Notice that these students are not left alone to their own devices to discover how to add. They are clear and secure in the tasks, understanding that what they are to do is reason, using what they know. The tasks are intentionally designed to lead to important place value, measurement, and number understandings and the skill of adding.

FREQUENTLY ASKED QUESTIONS

Q: How important is it that these tasks are real-world?

A: It is less important that the tasks are real-world and more important that the tasks are *truly problematic situations* and that they make the mathematics *realizable* (Fosnot & Dolk, 2001). In these tasks you see students intrigued, interested, seeking to find patterns to help them, and realizing important mathematical connections.

PROBLEM STRINGS

I've facilitated, written, and written about Problem Strings in Professional Development for years (see my *Numeracy Problem Strings* K–5 series; Harris, 2024). Problem Strings are an exceptional instructional routine that give us an enormous bang for our buck.

REFINING UNDERSTANDING OF PROBLEM STRINGS

You've already seen Problem Strings in action in earlier chapters in this book, but let's get clearer on what makes a string of problems into a Problem String. "A Problem String is an instructional routine that uses a set of related problems presented in a purposeful sequence to support students in developing visual, numeric, spatial, and operational relationships" (Harris, 2024, p. 8). Let's unpack that definition.

A SET OF RELATED PROBLEMS

The problems in a Problem String are related, not random. In one kind of string there are often helper problems that work in a sequence to help students solve more challenging problems. For example, in Numeracy Problem Strings: Second Grade, the second lesson on Remove to Ten (Harris, 2024, p. 84) starts with 13 – 3 and the following problem is 13 – 4. Putting these two problems in succession allows for a conversation about how you might use what you know about 13 – 3 = 10 and that 3 is included in 4, so removing one more than 3 will get us one lower than 10.

A PURPOSEFUL SEQUENCE

This string continues with two more pairs of related problems where removing to 10 helps with removing to just under 10: 16 – 6 is paired with 16 – 7, and 12 – 2 is paired with 12 – 3. Repetition of similar pairs of problems allows students to notice and make use of core elements, structure, and numerical relationships. Sequences of related problems interspersed with purposeful discussion allow for "combining discovery and copying. We take action to discover relationships, and when they are discovered, we recognize the similarity to something we might be able to copy" (Zull, 2011, p. 45). This is not *copy* as in *mimic* but copy as in *applying the same type of thinking in similar problems*.

TO SUPPORT STUDENTS

The purposeful sequence of related problems is only part of the power in this routine. Much of the power comes from the carefully crafted conversation led by the teacher. As they guide the discussion, the teacher constantly monitors who is applying the target strategy, who needs to be nudged to use new relationships, who needs to be further positioned as a valuable member of the mathematical community. All this influences who the teacher calls on to share, how many problems to use, the type of numbers to use, when to start putting words to the relationships and strategies, and when to cinch the learning with an anchor chart (Figure 8.1). All of these things work together to meet students where they are and help move them forward in their math-ing.

FIGURE 8.1 • A Sample Anchor Chart for the Get to a Friendly Number Strategy

Get to a Friendly Number

Keep one addend whole and break apart the other, so you can get to a friendly number

*partners of 10 are really helpful

56 + 27
(56 + 4) + 23 = 83

4 | 23
56 60 — 83

*partners of 100 are really helpful

75 + 27
(75 + 25) + 2 = 102

25 | 2
75 — 100 102

Source: Harris (2024), used with permission.

TRY IT

Facilitate a Problem String with your class. See Chapters 4 through 7 for strings to choose from.

The teacher constantly monitors who is applying the target strategy, who needs to be nudged to use new relationships, and who needs to be further positioned as a valuable member of the mathematical community.

DEVELOPING VISUAL, NUMERIC, SPATIAL, AND OPERATIONAL RELATIONSHIPS

After posing each problem, the teacher elicits answers and how students are thinking about the problem. Then, the teacher makes the thinking visible through models. This allows students to access each others' ideas. Modeling the thinking on the board creates access with a static representation that students can study and refer to. It also creates a record that can be used to compare to another student's strategy. In this string, strategies are modeled on open number lines, arranged one under the other so that it is easy to compare the relationships.

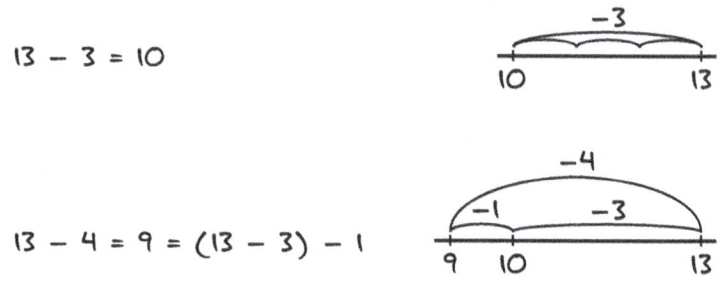

13 − 3 = 10

13 − 4 = 9 = (13 − 3) − 1

Source: Adapted from Harris (2024), used with permission.

Hearing students talk about their strategies is often not enough. Many students need a visual representation of the relationships so they can attend to the size of numbers, compare magnitudes, and make sense of multiple actions. I refer to using models as "making thinking visible," which also makes the thinking take-up-able.

MOVING THE MATH FORWARD

To promote student growth, Problem Strings must start with multiple access points—this means students with varying amounts of experience can access the first problem. As student thinking is represented and discussed, students at various experience levels can continue to access increasingly difficult problems.

During a string, the teacher is constantly monitoring where students are in the learning process: are students struggling, is the struggle productive, do students need more experience with certain relationships, are most students ready to put words to the target strategy, are students ready to model their own thinking, have students built multiple strategies enough that they are ready to compare strategies? The goal is to meet students where they are and help them build new relationships and strategies.

ANATOMY OF A PROBLEM STRING

The following is a general flow of a Problem String:

- The teacher asks an accessible first question of the class; students respond; the teacher represents the relationships on the board.

- The teacher asks the second question; the teacher circulates to find certain reasoning they want shared, the teacher elicits one or a few strategies, and models the thinking on the board.
- The string continues, repeating the process of pose the problem, think time, the teacher chooses who will share and then models the thinking on the board.

When enough students are using the target strategy, the teacher elicits student noticings about patterns. The conversation hangs here for a beat to support students in making connections about structures and relationships. Students who have not yet begun being metacognitive about the target strategy are looped into the conversation and given new ideas to consider. Then the problems resume with the intent that students can try using the new ideas that just came out in the discussion.

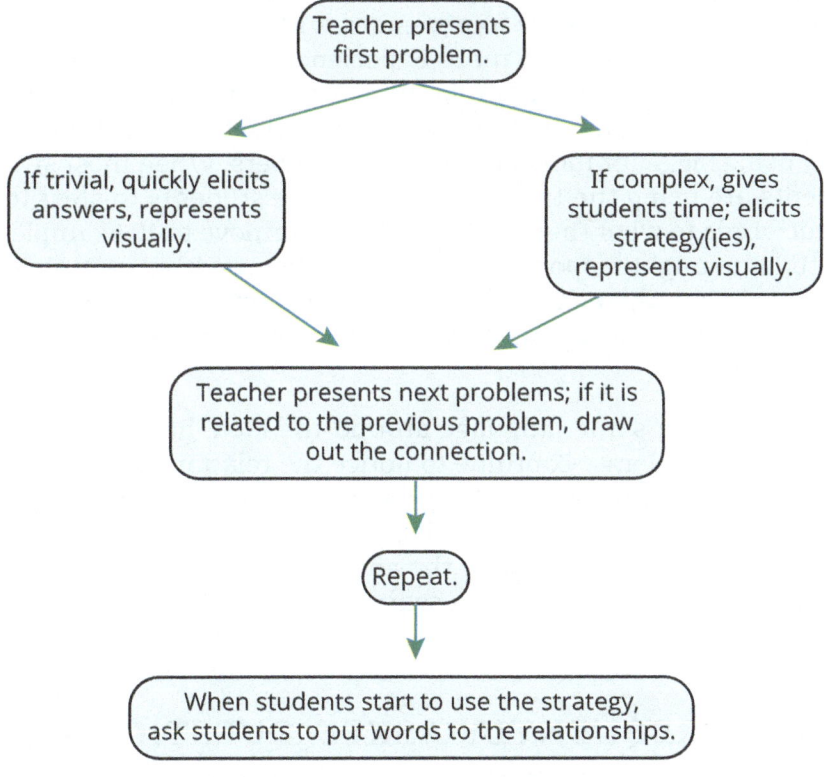

Source: From Harris (2024, p. 14). File used with permission

The teacher's job is to pull out ideas from students, focusing the models and conversation on ideas that the teacher knows will be useful to develop. The teacher looks for misinterpretations and misapplications and brings them before the community so that everyone can become clearer. Students start out grappling with relationships. Over the course of a string/series of strings, students gain clarity because of the conversation and their own experimenting using what they know.

The teacher looks for misinterpretations and misapplications and brings them before the community so that everyone can become clearer.

TIP
A clunker problem is a problem that is markedly more complex in some way than the previous problems in the same string.

Strings often end with clunker problems on which students can apply their new ideas. The Remove to 10 string ends with a final problem of 15 – 6. There is no helper problem of 15 – 5, thereby giving students the opportunity to create their own helper. This last step is key to helping students transfer the idea of using what they know about Remove to 10 to help with other subtraction problems outside of the Problem String.

Problem Strings are not intended to be a one-and-done routine. Students need multiple experiences with new ideas to solidify them. To give them more experience, you can use "echo" strings, strings that use similar numbers and work on the same big ideas. You can follow those with a string that ups the ante by using bigger or more complicated numbers. When most students are using the target strategy, invite students to start to put words to what they are doing. In the Remove to 10 example, students might respond with "It seems like we keep getting to 10, and then the next problem removes a bit more. Getting to 10 first is a friendly place to be, and then we can remove whatever is left."

Once you get some language around the idea, have students do more problems. Continue to notice the relationships. When enough students are getting consistent with the strategy, co-create an anchor chart with the class to name the strategy and show an example. Hang the anchor chart in the room, and refer students to it when necessary.

FREQUENTLY ASKED QUESTIONS

Q: When you say to co-create an anchor chart when students are getting consistent with the strategy, what do you mean? What does that look like?

A: Co-creating means you elicit student input. You have the plan of an anchor chart in mind and encourage dialog around the patterns students are noticing and using, steering toward a mathematical description. Let children choose a representative problem. Getting consistent means that the pattern (strategy) is occurring to them more frequently and they are able to make sense of how to use the strategy when it comes to mind because of the numbers involved.

Q: When you say that a teacher chooses who shares, how do I know who to choose?

A: Always choose with equity in mind. Be cognizant of positioning all students over time as helpful contributors. Look for students who are just starting to notice and use a pattern—often they will grapple aloud as they try to explain their reasoning. This is helpful because it gives the class something to discuss and clarify. It's rarely helpful to have the most clear explanation given; otherwise, you could just tell students everything. In a Problem String where you are developing a particular strategy, ask for a couple of different strategies at first. As the string progresses and as you do more strings for the same strategy, focus on representing the target strategy. Allow students to solve the problems however they want, but help them develop the target strategy by representing, discussing, and asking about that target strategy.

Q: Are Problem Strings the same as Problem (Number) Talks?

A: No. Problem Talks are a single problem that students solve, and the resulting conversation is about the many ways to solve it. There are a few reasons to use Problem Talks: to convince students that there are multiple ways to solve a given problem (you don't need this if you are doing Problem Strings), to assess which strategies students are using, and most importantly, to compare strategies for efficiency for a given problem. Students need many more Problem String experiences for every Problem Talk experience.

Q: Are Problem Strings the same as thin slicing (Liljedahl, 2021)?

A: No. Thin slicing is when students solve a group of problems with no teacher-facilitated class discussion between the problems. Problem Strings require teacher modeling of student thinking and discussion that highlights important relationships. Throughout a string, students have the ability to try ideas that are new to them because they have seen and heard the ideas from their community.

A GREAT TEACHER FIRST STEP

If you want to start teaching math as figure-out-able, start with Problem Strings. You don't have to adopt a new curriculum or overhaul your entire class to begin.

Facilitating one Problem String a week is a small, bite-off-able chunk to help students transition away from an answer-getting paradigm to a belief that math is figure-out-able.

There has been recent suggestion to use non-curricular tasks to get students thinking and then transfer those behaviors to mathematical content. My experience with using Problem Strings is that we can skip the non-curricular tasks and get students thinking and reasoning by jumping straight into Problem Strings. This is in part because a major tenet of participating in a Problem String is that *you use what you know*.

That sounds a little silly if you consider the alternative—*use what you don't know*. That's crazy, but that is often what we ask when math is seen as steps to memorize and mimic. During a string, students are encouraged to solve the problems anyway they can. Then, as we talk reasoning and students hear and see what others are doing, they gain more ideas than they had when the string began. The teacher keeps the lesson focused on building relationships, understanding each others' thinking, and striving for sophisticated, elegant solutions.

When you are ready to implement Problem Strings as more than just an instructional experiment, there are several things to think about. Problem Strings can be used before a Rich Task as a way of getting new ideas percolating and bubbling up. The big ideas are explored more fully in the Rich Task, and then Problem Strings are used to follow up and more fully develop relationships into strategies. The bulk of the schema building and refining is done through Problem Strings.

Problem Strings are also a great entry point for leaders and coaches because they can be used across all grade levels.

> Problem Strings are extremely helpful as a common structure that all teachers can use across all grades. Teachers at every grade level can have conversations about how they are choosing the model to use, which strategies they are having students share and in what order and why, how to decide whom to call on, how and when to anchor the learning, etc. It doesn't matter that the math content varies; all teachers can gain from discussing these choices. (Harris, 2025a, p. 238)

OTHER INSTRUCTIONAL ROUTINES

Besides Problem Strings, there are other important instructional routines. The following is not an exhaustive list, but I offer them as important high leverage activities on which to spend your already limited time.

I HAVE, YOU NEED

I Have, You Need is a quick and powerful instructional routine to help kids construct the very important numerical relations of partners of 5, 10, 20, 100, and beyond. I Have, You Need can be played with little to no material and very little planning. It can be done in small bits of time in class, on the go, and can be shared with families to work on at home.

To play, the caller (often the teacher) establishes a target total number and calls out a number smaller than the target. The partner (or other players) supply the amount needed so that, together, they reach the target amount. At first, the routine may be done in context, but it can quickly slip away as the routine progresses. Here's an example:

> You know I love having mints during the day! What if I said that I was going to bring 10 mints, and I was going to share some with one of you. If I eat 9 of the mints, how many would you need to eat so that we ate all 10?

When students know the amount needed to make a total of ten, they give a private signal until being cued to respond. The caller then suggests a different amount.

What if I have 7 of the _____? How many would you need so that we eat all 10?

What if I have 8? How many would you need?

I have _____, you need _____?

Between each question, the teacher observes students to gather information about who knows the partner amount, who is using a counting strategy to find the missing amount, and who is not able yet to find the missing number. This information helps the teacher make informed choices about future work.

> **TRY IT**
>
> Play I Have, You Need with your students. Start with numbers that are close to the goal. Gradually start to include commutative pairs back-to-back to support connections about partners of 10.

VARIATIONS ON I HAVE, YOU NEED

With early learners, a target goal of 5 is a great start!

Watch a great sequence with a young Hiram playing I Have, You Need.

qrs.ly/jjgl28s

If I have 3 fingers, how many to make 5?

With younger learners who own partners of 5, you might begin working with partners of ten on your fingers. Just like with partners of 5, you can flash the "I Have" amount and let students determine how many they need to total ten. Some students can imagine the fingers that are missing, while others will use the support of counting the bent fingers one at a time.

Another useful model to represent partners of ten is the number rack. As you announce, "I Have _____," slide beads to the left, leaving the "You Need" amount hidden.

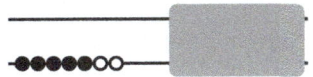

In this model, students cannot see the missing amount but can make use of what they already know about combinations of 5 within combinations of 10.

Because partners of 10 are crucial in early years, play I Have, You Need often. These partner relationships will help students with both addition and subtraction strategies. Look for suggestions about when to play I Have, You Need in each strategy section in Chapters 4 through 7.

Once students are familiar with partners of 10, they can use those partners in two distinct ways, both of which are important leaps in their learning journey (Table 8.1).

TABLE 8.1 • Making Important Connections

Students can use what they know about teen numbers and partners of 10 to make sense of partners of 20. To help make this connection, play I Have, You Need partners of 10, recording equations on the board or on a poster, in random order. Then, begin calling out teen numbers in the same order as recorded for partners of 10. In discussion, draw out the partners of 10 within each partner of 20 and how partners of 10 can be useful with partners of 20.	Partners of 10	Partners of 20
	7 + 3 = 10	(17 + 3) = 20
	5 + 5 = 10	(15 + 5) = 20
	6 + 4 = 10	(16 + 4) = 20
	8 + 2 = 10	(18 + 2) = 20
	2 + 8 = 10	(12 + 8) = 20
	9 + 1 = 10	(19 + 1) = 20
	1 + 9 = 10	(11 + 9) = 20
	3 + 7 = 10	(13 + 7) = 20
	4 + 6 = 10	(14 + 6) = 20

(Continued)

(Continued)

Students can also use partners of 10 to reason about partners of 100 for multiples of 10. Students can be nudged toward using what they know about ten with larger numbers. If 7 ones and 3 ones make 10 ones, then 7 tens and 3 tens make 10 tens, or 100.	Partners of 10 7 + 3 = 10 5 + 5 = 10 6 + 4 = 10 8 + 2 = 10 2 + 8 = 10 9 + 1 = 10 1 + 9 = 10 3 + 7 = 10 4 + 6 = 10	Partners of 10 70 + 30 = 100 50 + 50 = 100 60 + 40 = 100 80 + 20 = 100 20 + 80 = 100 90 + 10 = 100 10 + 90 = 100 30 + 70 = 100 40 + 60 = 100

Once students own partners of 100 for multiples of 10, a next move is to work with students to gain familiarity with partners of 100 for multiples of 5. Don't worry about fluency with *all* the partners of 100; that work will come in later grades.

This routine can also be played in pairs, where one student is the caller for another student; then they trade roles. Students may also play alone with a deck of premade cards to generate the number for them.

COUNT AROUNDS

A Count Around is a powerful instructional routine that can be used in many different grades to promote numeracy, especially place-value relationships and connections. The routine is fairly contained, easy to facilitate, and gives teachers a chance to hear student ideas and thinking.

Count Arounds begin with the teacher or students selecting a starting number for the class. The first student begins at that starting number and adds a specified amount. The class continues to count around, one student at a time, by that same specified amount.

For example, the teacher might choose to start with the number 8 and then tell the class to count around, adding 10 each time. The first student starts with 8 and adds 10 to get 18, the next student adds 10 to get 28, and so forth. As students count, the teacher writes the numbers on the board in such a way so that specific patterns emerge.

At certain, preplanned junctures the teacher stops the count and asks noticing, probing questions. In Count Arounds, the counting is not the most important part. Instead, it is the patterns that emerge and the discussion about the patterns. Figure 8.2 shows three sample Count Arounds in three different arrangements that might help students notice different patterns.

FIGURE 8.2 • Sample Count Arounds

Start with 8, count by 10	Start with 7, count by 9	Start with 5, count by 5
8	43	5 10 15 20
18	52	25 30 35 40
28	61	45 50 55 60
38	70	65 70 75 80
48	79	85 90 95 100
58	88	
68	97	
78	106	
88	115	
98	124	
108	133	
118	142	
128	151	
138	160	
148	169	
158	178	
168	187	
178	196	
188	205	
198	214	
208	223	
218	232	
228	241	
238	250	
248		

Start with 8, count by 10:
- What patterns do you see in the ones?
- What patterns do you see in the tens?
- Do the tens keep counting up? 17 tens in 178?

Start with 7, count by 9:
- What patterns do you see in the ones?
- What patterns do you see in the tens?
- Did anyone use the pattern to add 9?
- Where does that pattern break down? Why?
- Did anyone use +10 to help you with +9? How?

Start with 5, count by 5:
- What patterns do you see?
- What is happening in the tens places? The ones places?
- What change occurs in a row? A column? Vertically?

Chapter 8 • Tasks to Develop Mathematical Reasoning

> **TRY IT**
>
> Facilitate a Count Around with your students. What patterns will you help them notice?

ANALYZING STRATEGIES WITH SMUDGE PROBLEMS

One way to help students develop important mathematics relationships is to have them analyze strategies where you have strategically *smudged* out certain numbers. Students fill in the missing information and look for commonalities.

For example, if students are working on subtraction and on using 10 as a helpful landmark, it can be helpful to examine number lines that show 10 being used in subtraction problems and then craft conversations with students about what the strategies have in common.

Figure 8.3 shows part of a student page from my *Foundations for Strategies: Single Digit Addition & Subtraction* kit (Harris, 2025b).

FIGURE 8.3 • Analyzing Strategic Smudges

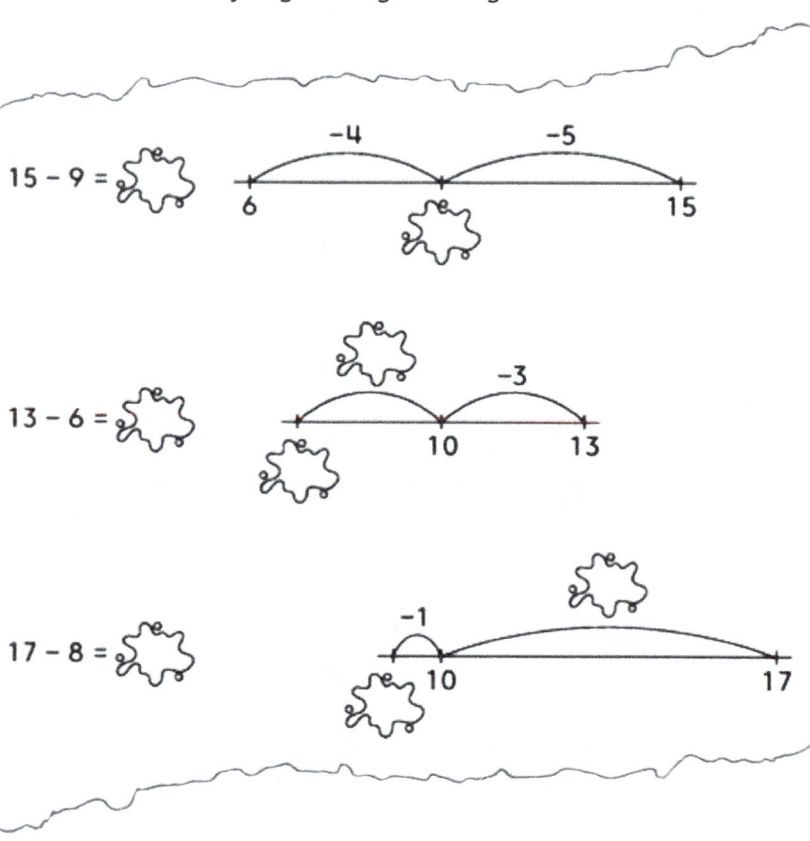

Source: Foundations for Strategies: Single Digit Addition & Subtraction, used with permission.

Students see three problems and their solutions, where there is a "spill" smudge in strategic places. When you fill in the smudges, what do you notice? For the first line, where a student is solving 15 – 9, how is that student using 10? How about the student solving 13 – 6? How does that compare to the students solving 17 – 8?

As they fill in the blanks, they are strengthening their subtraction skills, their sense of the importance of the number 10, and partners of 10 when they remove the leftovers. As students realize that each problem in Figure 8.3 is being solved by removing to 10 and then removing the rest, they are identifying the Remove to 10 strategy.

"ON THE FLY" ROUTINES

In many educational situations, there are bits of time that tend to go unused academically for a variety of reasons. You're waiting in line. Waiting for another group to leave. So much waiting. There are also bits of time where you want to squeeze something in but don't have quite enough time to assemble materials, adequately partner students or finish a whole lesson/activity/game. In these bits of time, "on the fly" routines are ideal. They can be done with little to no preparation, little to no material, and can be done anywhere.

These routines help students have needed experiences to be fluent with important mathematical relationships. And once fluent, these foundational relationships make strategic thinking even more accessible.

Keep these routines snappy and playful. Send the message that it's cool to play with these relationships—because it is!

> ### TRY IT
>
> Identify a bit of time in your day that you could try an on-the-fly routine. Choose one of the following activities to do during that time. Keep it snappy; make it fun!

MORE AND LESS

A super starting routine for very young students is "More." Simply choose a number and call out, "What's more than _____?" Students can give a private signal when they have a number in mind and reply with any number greater than the one called out. Students can share their number with a partner, whisper it aloud to the teacher simultaneously, or wait to be called on. Some students may think of the next number in sequence while others think of any greater number. After "More," "Less" can be introduced, where students think of a number less than the one announced. "More" and "Less" provide the opportunity for place value conversations, magnitude of number, greater than/less than, and work with hierarchical inclusion.

COUNTING BACK

Many young students can count back when starting with 10. Instead, begin the count with a single digit number less than 10. Instead of beginning with 20, start with a teen number. When you get to zero, say zero, not "blast-off!" At times, end the count before reaching zero.

+/– 10 OR 9

Pick a starting number between 10 and 20 (start with a teen number, then increase in magnitude over time, as appropriate). Ask, "What's 10 more than _____?" Allow students to figure 10 more in whatever way they can, then give a private signal when ready to share. Students can share their number with a partner, whisper it aloud to the caller simultaneously, or wait to be called on. Over time, note the pattern of the ones place staying the same each time, to nudge students who are still counting by ones. Ask "What's ten more" often for a variety of numbers. Then, move to including "What's 9 more?" as a second question.

The following is an example:

> "What's 10 more than 16?"
>
> "26!"
>
> "Ah, if 10 more is 26, then what's 9 more? Will that be more than 26 or less than 26?"

Work with students on 10 and 9 more until they are fluent, then introduce 10 less. After working with 10 less for some time and noticing patterns, introduce 9 less.

+/– 100 OR 99

To extend the work with plus or minus 10 and 9, choose any two-digit number, and repeat the routine with plus 100 for some time, then plus 99. Again, note aloud patterns students see and use when determining the sum.

DOUBLE

Knowing doubles is handy for solving unfamiliar problems, and we can build on students' intuition about doubles by doubling with them, a lot! Start with a small number and say, "What's double 3, 3 + 3, that's like 3 pairs of socks!" Let students count if needed. Work with doubles under 10, then move toward doubles under sum of 20. Use context as needed to help students imagine the doubling, and note aloud that patterns of moving from double 3 to double 4 means 2 more total (one more sock on each foot).

You could visually model student strategies, or because this is a quick routine that you want to do often, you can just have students share their strategies verbally. Ideally, you would do both, sometimes taking the time to model visually for all to see and sometimes just hearing their strategies.

HALVE

Once students can double, halve with them. Start with a familiar number like 10 or 20, and ask them to find half. Then ask even numbers under 10, then under 20. Extend halving to odd numbers with purposeful sequencing. "What's half of 8? What's half of 10? Ah, I wonder what half of 9 might be?" Using a context can be helpful: "If you and I are sharing 5 cookies, how many cookies do we each get? What will we do with that leftover cookie?" Connect halving a number like 12 to its double for addition, 6 + 6.

Each of these "on the fly" routines can be done orally, but you can also represent patterns on a small dry-erase board. These routines are great to share with families who want to support students at home to build relationships and reasoning.

FREQUENTLY ASKED QUESTIONS

Q: Why didn't my favorite routine make your list?

A: There are many excellent instructional routines out there. I think you will get more out of making a few important routines become *routine* by making them a priority. This means spending more time, especially at the beginning of the year so that students become familiar and comfortable with their role in these high leverage routines. If you have a few favorite others, feel free to bring them in when students are successfully making the most of the routines discussed here.

FREQUENTLY ASKED QUESTIONS

Q: Why didn't Number (Problem) Talks make your list?

A: At these grade levels, a good enough problem to spend substantial time on could be considered more of a Rich Task. If you are going to do a shorter, less involved problem, then do a string of them—that's a Problem String.

GAMES

Descriptions of a few of our favorite games.

https://qrs.ly/t9gl274

Well-chosen games can be a fantastic way for students to practice using the mathematical relationships they've gained to solve problems. Games can help students solve many problems in a short period of time in a playful way, with a low cost of failure. It's particularly nice if the games are self-checking, where students naturally see if their results are correct or not. Look for games that do not value quick retrieval over sophisticated reasoning.

At Math Is Figure-Out-Able, we've curated and created many games. We consider the best games those that can be continually played over time, upping the ante as students get better at them.

This allows you to take advantage of one explanation time that reaps benefits over many playing sessions. The best games also have open access to students, sometimes with just small tweaks, so everyone in the class is playing the same game but at their particular level of difficulty and challenge. We also like games that help build important relationships, not random fact retrieval.

FREQUENTLY ASKED QUESTIONS

Q: What do you think about computer/electronic math games?

A: I am aware of a few decent games that help students develop mathematical reasoning. At the same time, I am not a big fan of overarching programs that purport to provide individual instruction. No computer interface can replace the interaction with an expert teacher. Be wary of those programs that suggest they can. Skip apps that emphasize speed to get answers or remediate by telling students what to do as steps to mimic. Look for the end goals of apps—is it about developing mathematical reasoning or getting proficient at steps of an algorithm?

HINT CARDS

To help students automatize the addition facts, teachers can use personalized hint cards. These are just like addition fact flash cards with an added space for students to write in an individualized hint. In my *Foundations for Strategies: Single Digit Addition & Subtraction* (Harris, 2025b), we give teachers reusable, erasable sets of hint cards for each student.

3 + 7	5 + 6
7 + 3	6 + 5
Hint:	Hint: use 5 + 5

Source: Foundations for Strategies: Single Digit Addition & Subtraction, used with permission.

After students have been working to develop the mathematical relationships you've been reading about, they can be ready to work on facts they don't know yet.

First, interview students to find their individualized set of facts to work on. Show the student addition hint cards (without hints), one at a time. Make two piles: one of facts they know and one of those they do not know yet. If students answer readily, without undo effort and without counting by ones, that is a fact they know *well enough*. If they count by ones or put a lot of effort into figuring it out, that is a fact still to be worked on. You can ask students which facts they feel they would like to feel more confident with.

Then, help students create good hints for themselves for that fact, based on mathematical relationships, and have them write the hint on that card. Strive to write *just enough*. We don't want students to get bogged down in reading the hint, nor do we want the hint to be too obvious.

Ask students to work together to help each other. As students are quizzing each other in pairs, circulate and listen in. If a student seems to get stuck on a fact, check in. Ask how they are using that hint. Is it a good hint for

TIP

Remind students that their goal is to help their brain use the hints well so that the facts become automatic. This is not a race. It's not about how fast you are—it's about how well your brain uses the hint.

them? If not, reference the other strategies that they have been learning, and have them choose a new hint.

Check in with students over time. When they've learned a fact well enough, put it in their "facts I know" pile. Celebrate with them as their group of "facts I do not know yet" shrinks.

Of course we want students to know their facts! How they learn their facts is best done through developing additive relationships that will carry on to higher mathematics.

TIP
Once students are solid with their hints, send a copy of their personalized hint cards home.

Conclusion

Well-designed tasks, instructional routines, and games provide rich classroom experiences that optimize student learning. These include Problem Strings, the routine every classroom needs to high dose students with the major relationships that lead to the major strategies.

Sequence these lessons and activities in a way that acknowledges and integrates the varied paths of learning students travel. Cycle back as needed, always with an eye on the major landmarks and helping move the mathematics forward. Always keep in mind that your main goal is to help student realize that math is figure-out-able!

Discussion Questions

1. Discuss the Taking Inventory task sequence. Did anything surprise you? How do you think this sequence would work for your students?
2. Discuss the Measuring for the Art Show task sequence. Did anything surprise you? How do you think this sequence would work for your students?
3. What is an instructional routine? How does an instructional routine differ from other kinds of classroom routines?
4. How has your understanding of Problem Strings increased?
5. If the most important part of a Count Around is the discussion about the resulting patters, how can you leverage your professional community to help you notice important patterns?
6. How do you currently use games in your math time? Will that change? How?

CHAPTER 9

Modeling and Models

"Pam, please help. These parents are upset, and I'm not sure what to do," says Mary, a second-grade teacher at my children's school. She was one of the teachers working with me as we experimented with teaching real math. The parents of one of her students, Callie, had met with Mary to express their concerns about the way their daughter was drawing pictures instead of just solving problems the conventional way.

I knew what we (Mary and I) had been doing in my professional development sessions, but I wasn't sure what was happening with Callie. I called in the calvary—Kim was teaching third grade. She and I had a conversation about the situation, and she said, "Let me talk to Callie. I bet I can figure out what's up." We arrange for Kim to assess Callie while I could be there.

Walking in the door, it is obvious Callie is nervous. Before she sits down, Kim greets her and says, "Hey, before we get started, could you help me with something quickly?" Callie immediately relaxes and dives in to *help* Kim. In the midst of *helping*, Callie starts to add 27 and 14. She draws a number line and starts at 27. She says aloud, "I think I'll jump by 2s." She draws several jumps.

Kim politely interrupts, "Why 2s?"

"I like to jump by 2s," Callie replies as she continues drawing jumps. "We're supposed to make jumps." She draws jump after jump, not counting them.

Kim says, "Let's look back here. You started at 27?"

Callie nods.

"Where do you land if you make a jump of 2?" Kim asks.

Callie thinks for a moment and answers, "29. But I'm going to finish making these jumps." And she keeps drawing jumps.

Kim smiles. "Hmm, 29. That doesn't seem very friendly. Do you know any numbers near 27 that are friendly?" This discreet question draws attention to something Callie ideally would have been focusing on all along: *why* she was making a jump in the first place.

> **TIP**
>
> "We're supposed to" is almost always a warning sign if given as a justification for how a student is solving a problem. It suggests that the student is trapped in mimicking land instead of growing their reasoning.

No step taken as part of a strategy should ever feel random. No step of a strategy should ever be taken without a clear purpose in mind. Otherwise it's not a strategy.

Callie takes another thoughtful pause, looks at Kim, and says, "30, that's close and friendly."

Kim takes a pencil, draws a new number line with 27, and asks, "What jump gets you to that friendly 30?"

$$\overset{\;\;|\;}{27}$$

When Callie answers, "3", Kim hands Callie the pencil. "Make that jump. Yeah. Now, you've added 3 and you're at 30. How much were you supposed to add?"

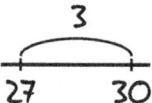

Callie answers, "14," so Kim replies, "And you've already added 3. How much more do you need to add?"

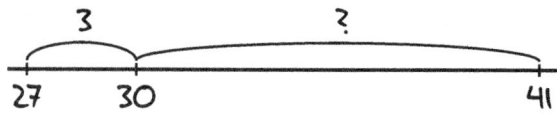

"11. I'd have 11 left. That's 41."

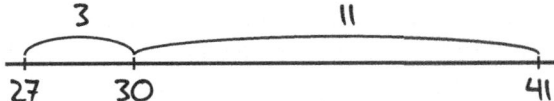

Kim nods. "Hey, I'm kind of curious. Back here at 27, do you know what 27 and 10 is?"

When Callie answers, "Yes, that's 37," Kim draws a new number line underneath the first with the jump of 10.

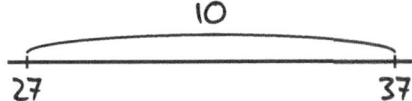

Kim continues. "But you were supposed to add 14. You've added..." and trailed off.

"10 so I need to add 4 more." Her eyes got wide with excitement. "And 37 and 4 is also 41!"

Kim later summed it up to me this way: "It's not about random jumping. It's purposeful moves. It matters where you land."

Of course Callie's parents were concerned. They were seeing random jumping, with lots of small jumps to solve problems. Why do all that picture drawing when they sensed Callie could solve the problem with just a few steps with an algorithm? Callie's focus seemed to be on drawing jumps because she thought that was the goal, not on thinking or finding the answer in an efficient way.

This shift, from a goal of "do the model" to a goal of "solve a problem in a clever and efficient way, assisted by the model," is absolutely critical. As our experience with Callie demonstrates, the difference between the two will not always be obvious to students or to the teachers teaching them. Given that reality, effective uses of models and modeling to develop math-ing requires deliberate teacher moves rooted in a thorough understanding of the content paired with a sound evaluation of where a given student's understanding is.

> *This shift, from a goal of "do the model" to a goal of "solve a problem in a clever and efficient way, assisted by the model," is absolutely critical.*

Notice that Kim's first move, after creating a safe learning environment, was to evaluate where Callie was, not to jump to conclusions based on secondhand reports. Callie's parents didn't have a problem with clever problem solving; they had a problem with incorrectly implemented models masquerading as clever problem solving.

Kim focused on two of the major strategies, Get to a Friendly Number and Add a Friendly Number. She used a model to help keep track of the relationships they were using. This is the purposeful use of models and modeling based on the major strategies to help students build and use major relationships. This chapter helps illustrate the use of models and modeling to develop math-ing.

But first, let's get even more clear on some terms.

STRATEGIES VERSUS MODELS

Strategies and models are different. A *strategy* is how you deal with the numbers to solve a problem. A *model* is a representation of a strategy, of relationships. It is important for teachers to get this difference straight.

Strategy	**Model**
how you deal with the numbers or structure to solve a problem	representation of a strategy, of relationships; some models can be tools

Source: Harris, 2024, The Most Important Numeracy Strategies, used with permission.

In Chapters 4 through 7, you learned about the major strategies, like Get to 10 and Remove a Friendly Number Over. To make those relationships visual, the strategies were represented on models, like a number rack or open number line.

Why is it so important to be clear on the differences between models and strategies? Notice how clarifying helps Mia and me focus on her thinking.

"How do you know?" I ask Mia.

Mia replies, "I used my fingers."

"How did you use your fingers?" I press for the strategy because she had told me a model, fingers.

Mia shows that she counted on. Her strategy is Counting On. Her model is fingers.

If a teacher is clear on strategies and models, then students tend to focus on their strategies and use models to represent their thinking and as tools to solve problems.

If teachers are not clear, everyone is muddy.

> **TIP**
>
> When students tell you how they know by saying the model, "I used cubes," "I did a number line," or "I did a jump strategy," reply with a version of "How did you use (that model)?"
>
> For example, when a student says, "I did a number line," you can reply, "How did you use a number line?" and, if needed, "Where did you start?" or "What was your first jump?"
>
> Questions like these will get at the relationships the student is using. That's their strategy.

FREQUENTLY ASKED QUESTIONS

Q: Do you mean general problem-solving strategies, like solve a simpler problem?

A: No, when I am referring to major strategies, I mean using certain important mathematical relationships, *like getting to 10 and adding the rest* or *subtracting too much and adjusting*. I also do not mean teaching strategies, like questioning, partner talk, formative assessment, or classroom management. Those teaching strategies are important, but in this book I'll call those teacher moves. I'll reserve *strategies* for the way we use mathematical relationships to reason through a problem.

We see this confusion often when teachers do Problem Talks (Number Talks), calling on any student who volunteers, and then putting their *method* on the board, without regard to whether it's a strategy or a model. If the goal is to hear from all the voices in the room, this might seem like a good idea. But I would argue that there are better ways to hear from all students and move the math forward *at the same time*.

Chapter 9 • Modeling and Models

There may also be a time where a teacher's goal is to highlight that a single strategy can be represented on more than one model. This is a less frequent and short-lived goal. Consider the following: A teacher might represent a student who used their fingers, starting at 5 and added 3 more, 5, 6, 7, 8, by drawing one hand up and then three more fingers on another hand. Then when this teacher calls on a student who used counters, the teacher might draw a big 5 and then 3 more counters. And when another student says they used tally marks, the teacher writes that. All of these are the same strategy, just different models.

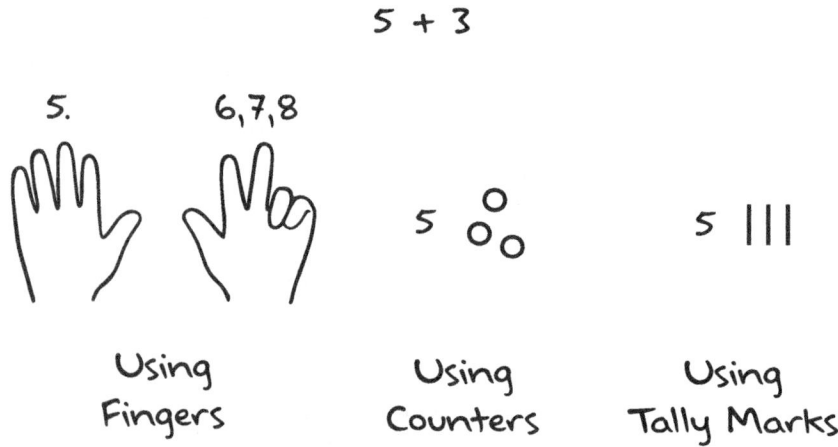

Three Models Showing the Same Counting On Strategy for 5 + 3

A more important and longer lasting goal is to highlight and compare *strategies*, not *models*.

If your purpose is to compare strategies in a Problem Talk for 5 + 3, look for students who are Counting Three Times and others who are Counting On. Have one student share who used each of these strategies (Figure 9.1). If you've already represented Counting Three Times with fingers and a student shares Counting Three Times with counters, you could say, "Right, did you also count out 5, count out 3, then count them all? Yes, and you did it with counters? Great. Let's find someone who kept the 5 whole, who started with 5. Danny, I think I saw you doing that. Will you share your thinking?" In other words, comment that the same strategy could be shown/done on a different model, but don't represent it. Look for a different strategy to represent.

FIGURE 9.1 • Anchor Chart Comparing Counting Three Times and Counting On

Access the QR code for more help parsing out the difference between models and strategies found in *Major Numeracy Strategies* (Harris, 2024).

https://qrs.ly/r6gl277

Similarly, if you are comparing strategies in a Problem Talk like 38 + 29, you are looking for different strategies, not different models. Figure 9.2 shows a sample display. When students share their thinking, you could choose to use an open number line to represent the thinking even if a student does not talk about jumps.

FIGURE 9.2 • Sample Display for a Problem Talk Comparing Strategies

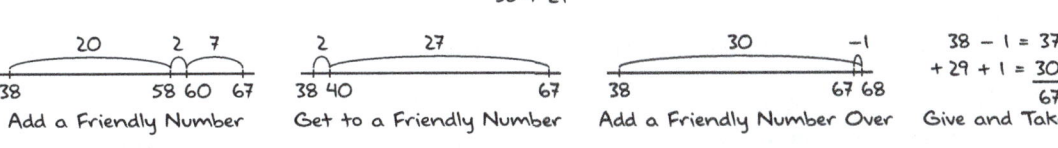

FREQUENTLY ASKED QUESTIONS

Q: Why are you making such a big deal about the difference between model and strategy?

A: Because if we are not clear on the difference between a model and a mathematical strategy, we cannot focus on developing more sophisticated strategies. If students are not getting more sophisticated in their strategies, they will not be efficient enough, and teachers will unnecessarily feel forced to teach mimicking algorithms.

(Continued)

> (Continued)
>
> **Q:** If strategies are so important, why do you have a whole chapter on models?
>
> **A:** Because models help make thinking visible, which helps students create mental mathematical relationship maps. These models also become tools to help students reason using those relationships. Read on to find out how!

THE MANY MEANINGS OF MODEL

There are many ways we are currently using the words *models* and *modeling* in mathematics education. Because of this, much of the conversation can be confusing as we talk past each other, each using a different meaning and sometimes more than one meaning in the same sentence.

Let's parse out these meanings so we can focus on the most important ones.

MEANINGS OF MODEL NOT USED IN THIS BOOK

Table 9.1 illustrates three common usages of the word *model* that differ from the definitions used throughout this book.

TABLE 9.1 • Three Common Uses of the Word *Model*

Model (verb): to demonstrate, show how something is done	Model-Demonstrate
Since I do not advocate algorithm mimicking in math class, when you read *model* in this book, it will not mean "to demonstrate."	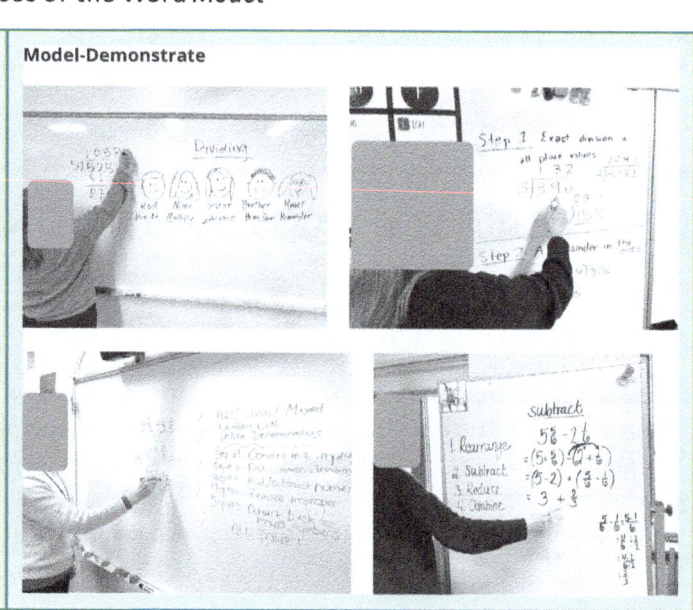

Models (noun): manipulatives	**Model-Manipulative**
When you read the word *model* in this book, don't think manipulatives unless I'm talking about manipulatives as a model to build relationships.	
Model (noun): a process	**Model-Process**
You may have used the modeling process when working with data or contextual problems	Problem → Formulate ↔ Validate → Report
The modeling process is great for students to use as they dig into really rich tasks, but that is not the part of models that I focus on in this book.	Formulate → Compute → Interpret → Validate

MEANINGS OF MODEL USED IN THIS BOOK

Throughout this book, my usage of the word *model* aligns with the central premise: whether employed as a verb or noun, *model* should fundamentally relate to the thinking process involved.

MODEL (VERB OR NOUN): REPRESENT THINKING OR A REPRESENTATION OF THINKING

The *representing thinking* (model as a verb) meaning is to make thinking visible, point-at-able, and discuss-able.

> We represent thinking to make a representation of thinking
>   (verb)   (noun)

This kind of modeling creates a representation of thinking (a model) for the purpose of examining the thinking itself. To model (represent) student thinking, we listen and ask clarifying questions as we use a visual and often spatial model to record those relationships.

A number line model can be used to represent student thinking: "I found 15 – 8 by going back 5, that's 10. Back 3 more, that's 7."	*number line showing jumps of −3 and −5 from 15 to 10 to 7*
Moving beads on a number rack or putting up fingers one at a time are examples of dynamic representations of thinking.	*bead rack and hands showing finger counting*
More permanent examples of representations of thinking are sketches of counters, open number lines, and equations.	*counters, ten-frame dots, open number line with jumps of 2 and 2 from 8 to 10 to 12* $8 + 4 = 8 + 2 + 2 = 10 + 2 = 12$

TRY IT

How might you represent a student who says, "I found 8 + 7 by starting with 8, adding 2 to get to 10, and then adding the rest"?

TRY IT

How might you represent a student who says, "I found 38 + 15 by starting with 38, adding 10 to get to 48, then adding 5 by adding 2 to get to 50 and then the leftover 3"?

These representations are a way of helping students realize that when their brain thinks a certain way, those relationships can be shown visually—"When I think that way, it could look like that!" As they compare their mental maps with the visual images, students get a chance to clarify and solidify their thinking. This can make their mental maps stronger.

TRY IT

Create a representation of a student's thinking who is solving 6 + 3 who says, "I know 6. Then 7, 8, 9."

> **TRY IT**
>
> Create a representation of a student's thinking who is solving 42 – 27 who says, "I subtracted 30 and landed on 12. But I subtracted too much so I added back 3 for 15."

MODEL (NOUN): REPRESENTATION OF THE SITUATION

A *representation of the situation* is a model of the relationship happening in the problem.

A K–2 example is number bonds. These show how the numbers are connected.

For a problem like 5 + 3 and 9 – 2, number bonds might look like the example that follows:

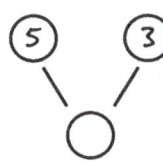
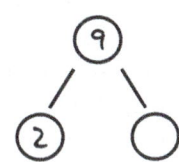

Although number bonds show the relationship between the numbers in the problem, they have little utility beyond that.

> **FREQUENTLY ASKED QUESTIONS**
>
> **Q:** You do not have any number bonds in this book. Why not?
>
> **A:** Students either do not need number bonds because they are already clear on the problem or students get stuck putting all their efforts into creating the number bond and never actually solve the problem. Number bonds are just not helpful enough. The other models in this book work better.

More helpful models of a situation are those that create a visual or spatial representation of a scenario or phenomena that can then develop into a tool for thinking, such as those discussed in Chapter 8.

MODEL: TOOL FOR THINKING

There is a small set of models that are excellent as *tools for thinking*. In other words, these models can become tools students can use *in order to think* (Fosnot & Dolk, 2001; Gravemeijer, 1999).

These include number racks (rekenreks), open number lines, and equations. You will read more in this chapter about these tools and how they can be used as *tools for thinking*. You have also seen them at work in the Problem Strings and the Rich Tasks in Chapter 8.

Now that we've defined *model* and *modeling*, let's connect to your prior experiences and tasks you've read about in this book.

PUTTING MODELS AND MODELING IN CONTEXT

Throughout this book, you've read about Problem Strings, lessons, and tasks you can use to develop mathematical reasoning. Let's parse out where and how the different definitions of models and modeling were and were not used in those teaching activities.

In this book, you have seen the use of manipulatives like cubes, number racks, and bundles of items, as well as models like number lines. One of the things that was initially so eye opening to me and profoundly influenced me is the way that the teachers Jodi (Taking Inventory) and Hildy (Measuring for the Art Show) from Chapter 8 used manipulatives and open number lines. These teachers were not using manipulatives to mimic steps of a procedure. They were also not using manipulatives to show students the math concepts, assuming that if students see the manipulatives, they will understand the underlying concepts. We do not hear Hildy say, "Use a jump strategy," as if a model is a strategy. These teachers used manipulatives to help students notice important patterns and realize the mathematics happening. They used number lines as tools for reasoning to build and use mental maps.

The distinction is subtle. And very important.

Remember Jodi and her students who were taking classroom inventory of things, like books? They counted the total number of books by ones, creating and counting the packs of 10, and noting how many books were loose.

You may have read that investigation, and based on your experience, thought, "Right, that's what we do in our calendar time. We keep track of how many days we've been in school. Each day we add a popsicle stick and, as needed, we bundle popsicle sticks into groups of 10 and leftover popsicle sticks. We call those bundles 10s and the leftovers 1s. We're teaching place value. Children recognize that the bundles are groups of 10s and that's how numbers are created."

There are several key differences between this calendar scenario and the Rich Task Jodi facilitated for her students. The calendar scenario describes a series of things to do: one more day on the calendar, one more popsicle stick, bundle together and call it a 10. Most students are not performing mental actions here. What there is to figure out, or why it needs to be figured out, is not clear. Do students understand place value after being shown this procedure day after day? Some may; most do not.

Consider how different Jodi's students' experience was. They counted large amounts, getting a feel for quantity, and because the numbers were large, sometimes they lost track. They were motivated to find ways to keep track. Subtle teacher moves suggested it might be easier to keep track if they put the items in groups, and groups of 10 sounded good. They kept track for a purpose that makes sense to them—taking inventory of classroom things so they would know how many paper bags they had left to make puppets. Jodi represented student thinking as she cleverly made a poster that had the potential to high-dose students with place-value patterns of 10s and 1s and the total. Students didn't stop at "ten 10s make 100." They were looking at 24 tens making 240. They stayed in context as they noticed the patterns between the 249 total and 24 packs and 9 loose. They were intrigued by the patterns and wanted to know if it would always happen.

> **TIP**
> The important part is not which manipulative to use but how to use manipulatives. Rather than tools for mimicking, use them as tool for reasoning.

Sometimes they lose track. They are motivated to find ways to keep track.

Remember Hildy and her students measuring papers for labels for their art show? You may have read about that exploration and been reminded of using cubes (*manipulatives*) to add and subtract, thinking, "Yes, we should use cubes to add and subtract. For 28 + 17, count out the 28 cubes (2 rods, 8 cubes), and 17 cubes (1 rod, 7 cubes). Then bring the cubes together, and put the rods together. Count the individual cubes, trading 10 of them for another rod, leaving 5 cubes behind. Then count the result, 4 rods and 5 cubes. Write those digits down, 4 and 5, and then say the answer is 45." This describes *a series of steps to do*. Students using manipulatives this way to mimic a procedure are not using the mental actions of a reasoner. Even though the medium this time is cubes rather than digits, students are still experiencing addition as repeating steps they've been shown.

On the other hand, cubes can be used to support student development. Hildy's students used cubes when they were measuring for the art show. Because they were cleverly given just 2 colors of cubes and they saw an alternating pattern of 5s hanging on the board, they used those alternating groups of 5 cubes to locate needed measurements. Lengths of papers become locations on a cube number line. Students were curious about and could reason using the alternating 5s in different ways, thinking about where to find and mark 14 or 36 or 42 on the corresponding paper strip. They were reasoning, using what they knew about landmarks to find those numbers just up or down from those landmarks. The paper strip is *a model of the situation*, a record of measurements.

When Hildy modeled student thinking about addition problems on a number line, she made student reasoning visible, a *representation of their thinking*. Students tried to verbalize the way they were attacking an addition problem (*their strategy*), and Hildy drew jumps showing whether they chose to add a friendly number or get to a friendly number. The cube number line and paper strip both transitioned to become an open number line, a mental map built in students' minds of the linear relationships between the numbers. Students could reason that to add a paper of length 21 cubes to a paper of 19 cubes, they could add 1 cube to the 19, leaving 20 + 20. They had agency, choosing to use relationships that made sense to them. They were intrigued to try to find clever ways. They were using the open number line as *a tool for thinking*.

They had agency, choosing to use relationships that made sense to them.

FREQUENTLY ASKED QUESTIONS

Q: How is using a rekenrek (number rack) to solve problems different from using cubes to repeat steps?

A: In the Problem Strings you've read about in this book, the teachers use a number rack to represent student strategies that help to advance reasoning. Teachers and students are using the 5- and 10-structure of the rack to build mental maps of the relationships of numbers to 5 and 10. As a student talks about the way they are seeing the numbers, the teacher moves the beads to make that thinking visible. This is using a number rack as a *representation of thinking*. Students can then envision the rack to use those maps to reason logically, solving problems using the rack. This is using a number rack as a *tool for reasoning*.

As you think about using manipulatives and models like number racks and open number lines in your teaching, here are some principles to keep in mind:

- Rather than using manipulatives to demonstrate a procedure, hoping that showing a procedure on a model will help students see the underlying concepts, instead use manipulatives to help students understand contexts, creating a model of the situation.
- Rather than asking students to use manipulatives to mimic steps to get answers to problems, instead ask students to use manipulatives to notice patterns, make sense of relationships, and use those connections to reason to solve problems.
- Rather than asking students to "do a number line strategy," (because this confuses models and strategies), instead ask students *how* they are using number lines, focusing on the way they are using relationships
- Use models like number racks and open number lines to represent student thinking, making their reasoning visible, point-at-able, and discussable.
- Ask students to represent their thinking using those models, once they have had their thinking modeled enough to be able to do on their own.
- Help students transition from using those models *representing their thinking* to using those models, number racks and open number lines, as *tools for reasoning*.

> **TRY IT**
>
> Take stock of the manipulatives in your classroom. Have you used any of them to get students to mimic steps of procedures? If so, which manipulatives could you shift to using to support reasoning and sensemaking?

EXPLORING MODELS BY THEIR BEST USES

Figure 9.3 shows examples of models that you will see being used in Kindergarten through second-grade classes. Let's discuss these models based on their optimal use cases:

- Building relationships
- Building relationships AND tools for computing

FIGURE 9.3 • Common Models in K–2

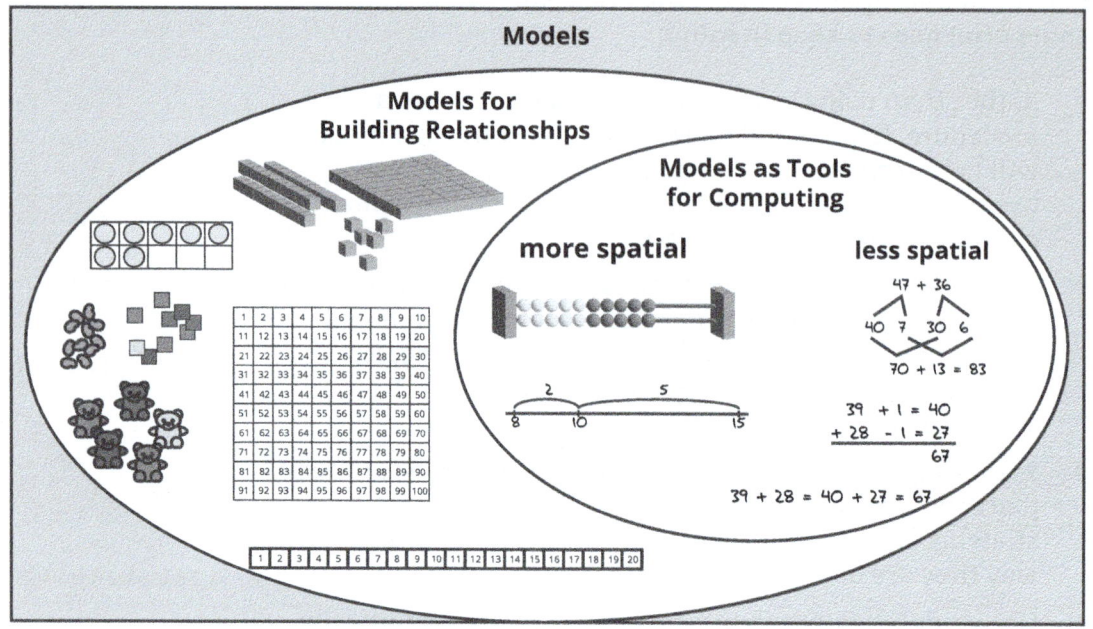

MODELS FOR BUILDING RELATIONSHIPS

Some models are effective in helping students develop numerical and spatial relationships. In the following descriptions, keep in mind that these models are not well suited to be used as tools for computing.

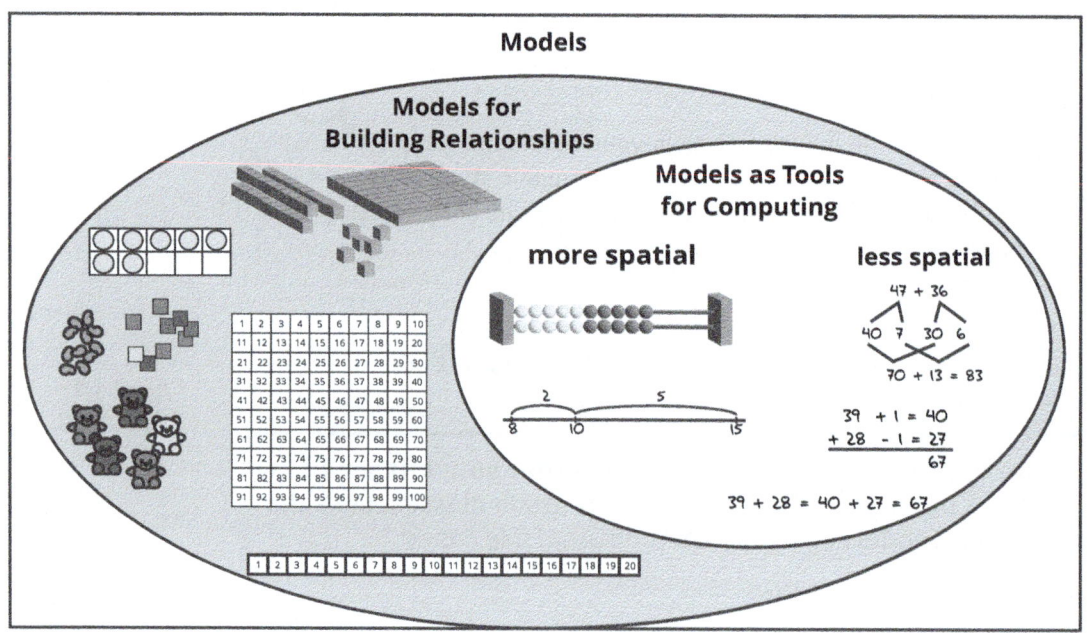

ONE-TO-ONE COUNTERS

Anything that suggests counting one by one is a one-to-one manipulative/counter. This includes beads, blocks, chips, teddy bears, fingers, and tally marks. Fingers and tally marks have the added advantage of grouping the single counts into groups of 5.

One-to-one counters are fantastic models for children who are learning to count. Once children begin to own most of the counting principles, it's time to move to models that suggest the 5- and 10-structure. To begin, that means shifting from tiles and beans to fingers and tally marks, and then to 10-frames and number racks.

10-FRAMES

A 10-frame is a simple 2 by 5 table where students fill in a number of dots to represent an amount.

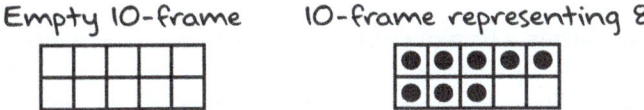

The brilliance of a 10-frame is the suggestion of the 5- and 10-structure for numbers within 10. Because of the 2 rows of 5 squares, numbers can be represented in their relationship to 5 and/or to 10.

For example, the number 4 is 4 filled in squares, but it is also 1 less than 5.

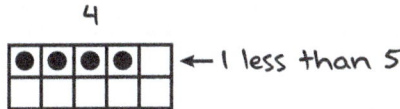

The number 7 is 2 more than 5 (because the whole row if filled in) and also 3 less than 10 because 3 are not filled in.

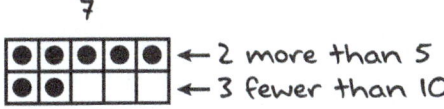

Use 10-frames while students are learning to count and learning to solve problems using Counting Strategies.

Although 10-frames are great to use while students are learning to count and solve problems using Counting Strategies, do not use 10-frames to compute. This use of the model encourages students to count one by one. Notice that to solve this problem using 10-frames, students must first count out 5 to fill the first frame, then count out 4 to fill the second frame, and then count those 4 one at a time as they move them to fill a frame.

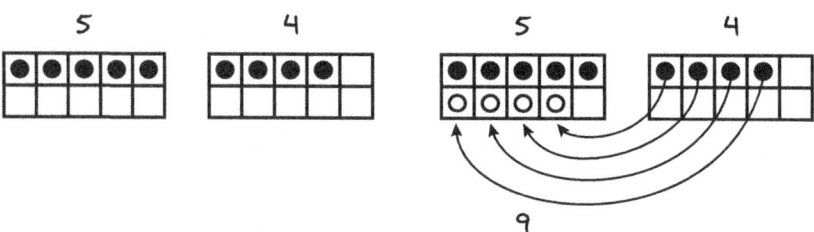

That's too much one-by-one counting when the goal is to build Additive Reasoning.

NUMBER PATH

A number path is a line of squares, each with a counting number.

1	2	3	4	5	6	7	8	9	10	11	12	13	14	15	16	17	18	19	20

This model can be helpful for students who are learning the counting sequence or the look of a numeral. Because they require one-to-one counting, do not use them to compute.

HUNDREDS CHARTS

A hundreds chart is a version of a number path, where the structure of numbers from 1 to 100 or 0 to 99 is represented.

1	2	3	4	5	6	7	8	9	10
11	12	13	14	15	16	17	18	19	20
21	22	23	24	25	26	27	28	29	30
31	32	33	34	35	36	37	38	39	40
41	42	43	44	45	46	47	48	49	50
51	52	53	54	55	56	57	58	59	60
61	62	63	64	65	66	67	68	69	70
71	72	73	74	75	76	77	78	79	80
81	82	83	84	85	86	87	88	89	90
91	92	93	94	95	96	97	98	99	100

0	1	2	3	4	5	6	7	8	9
10	11	12	13	14	15	16	17	18	19
20	21	22	23	24	25	26	27	28	29
30	31	32	33	34	35	36	37	38	39
40	41	42	43	44	45	46	47	48	49
50	51	52	53	54	55	56	57	58	59
60	61	62	63	64	65	66	67	68	69
70	71	72	73	74	75	76	77	78	79
80	81	82	83	84	85	86	87	88	89
90	91	92	93	94	95	96	97	98	99

Hundreds charts are fine models with which to build relationships, but don't use hundreds charts to compute. To use them to add or subtract, you end up using too much counting by ones. When developing Additive Reasoning, we don't want all that one-by-one counting.

BASE 10 MATERIALS

Base 10 materials are structured to use volume to represent our base 10 number system. There is a 1-unit cube, a row of 10 cubes to represent 10, a flat of 100 cubes that represents 100, and a cube of 1000 units that represents 1000.

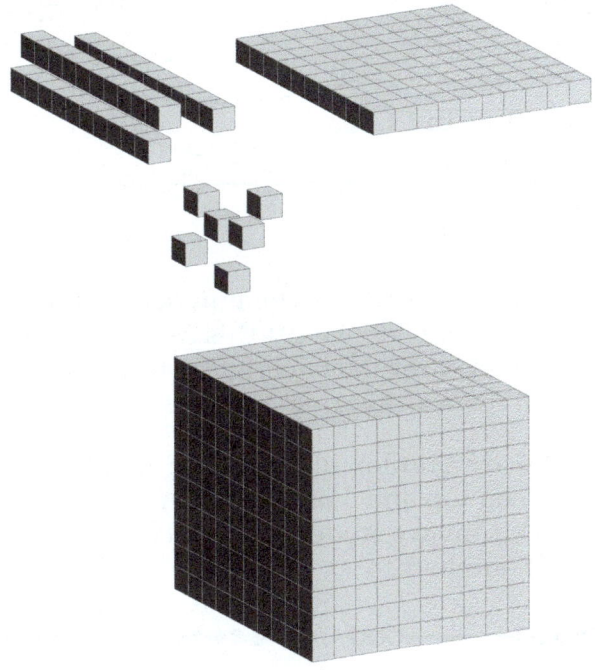

These are fine to build relationships, but because they require a ton of one-by-one counting, they are not good models with which to compute.

MODELS FOR BUILDING BOTH RELATIONSHIPS AND COMPUTING

There is a small set of models that are excellent both for building relationships and as tools for computing.

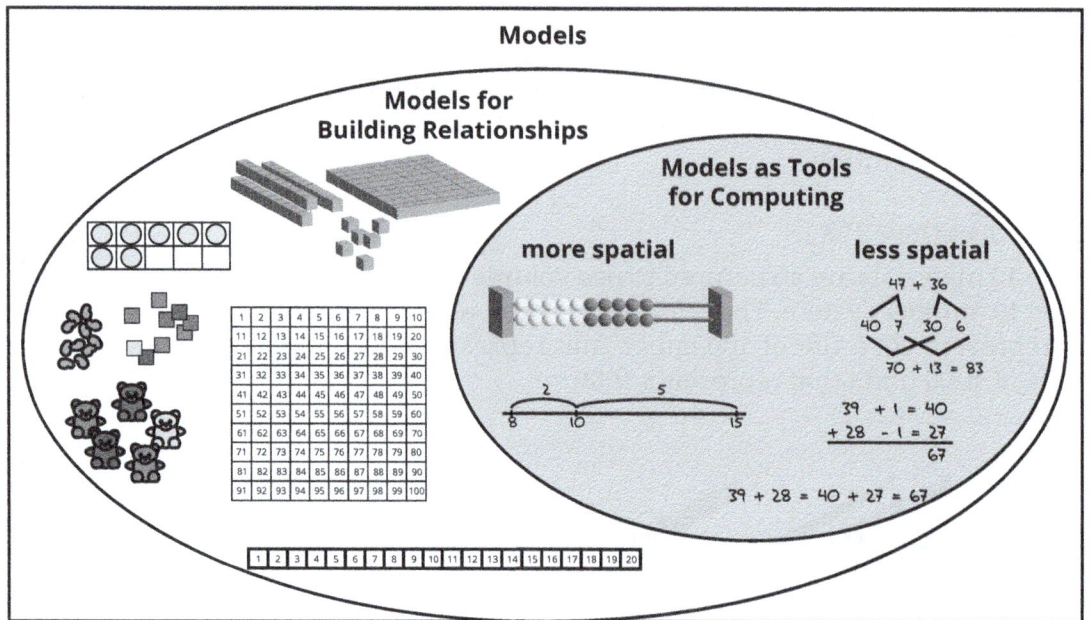

NUMBER RACKS

Number Racks are fantastic tools for number sense and addition and subtraction within 20. We get this model from the Dutch, who named it a rekenrek. More correctly translated to English would be to call them a "figuring" rack.

Number racks are similar in structure to 10-frames, with 10 red beads, 5 on top and 5 on bottom and the same with 10 white beads. This invites students to notice and use the 5- and 10-structure of numbers to 20.

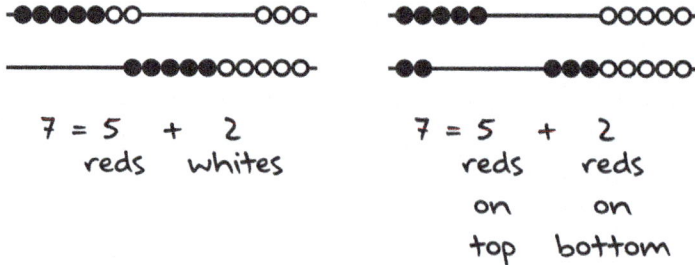

Once students are successfully counting to the teens, transition from 10-frames to the more versatile number rack. "Because a number rack allows for movement of more than one bead at a time, it is a better tool for modeling student thinking and for computing than a static 10-frame" (Harris, 2025, p. 3).

Start the number rack with all beads on the right and move beads to the left to represent the problem. "As you represent student thinking, slide beads slow *enough* for students to track your movement. Resist the urge to show *your* thinking—instead, represent what students say" (Harris, 2025, p. 3).

Move beads in groups, with as few pushes as possible, not one bead at a time. Remember, you're building Additive Reasoning.

This model can be used to represent both meanings of subtraction: difference/distance and removal.

Notice that in Figure 9.4, 12 beads are shown on the left. You can represent the difference between 12 and 2 by asking, "What is the difference between these 12 and just the 2 on the bottom? Sure enough, just those 10 beads."

FIGURE 9.4 • Representing 12 − 2

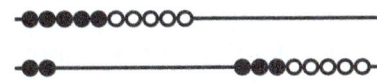

Watch to see Kim as she introduces a number rack to a first-grade class.

https://qrs.ly/bxgl278

You can also use the 12 beads on the left to represent removal, "Here are 12 beads. How many are left when you remove 2? Right, just those 10 on the top."

Introducing the Number Rack to Students

Just like with any manipulative, first give students a chance to explore it before expecting students to do anything with it.

THE OPEN NUMBER LINE

Whereas a number path is a path of discrete squares, a number line is a continuous measurement model.

Discrete means distinct, separate values. These values can be counted, and there are no gaps between values. Examples include the number of students in your class and the number of teddy bear counters in the pile. Continuous data are unbroken or uninterrupted and can be measured with infinite precision. You can find any value within a certain range. Examples include the length of your pencil or how tall you are.

A closed number line shows all the tick marks in a range, while an open number line is a more versatile tool because you put tick marks where they are helpful. Figure 9.5 shows the problem 7 + 5 with a Get to 10 strategy on both a closed and an open number line.

FIGURE 9.5 • Closed Versus Open Number Lines

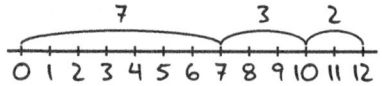
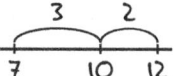

Representing 7 + 5 = 7 + 3 + 2 = 10 + 2 = 12

Closed number lines can be tricky models for students to compute with because students can be still grappling with the discrete tick marks versus the continuous nature of the line. Because of this confusion, and because closed number lines can encourage one-by-one counting, use closed number lines to help build relationships but not as tools for computing. Open number lines, once well developed as a mental map in students' brains, become a wonderful model to build relationships and as a tool for computing.

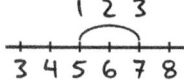

Is 5 + 3 = 7?
Do I count the tick marks?

5 + 3 = 8
Do I count the span?

Difference/Distance and Removal

The open number line model can be used to represent both the strategy of finding the difference/distance and removal strategies to solve subtraction problems.

17 − 8

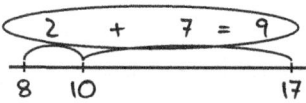

The difference between
8 and 17 is 9.

17 remove 8 lands on 9.

52 − 48 = 4

52 − 7 = 45

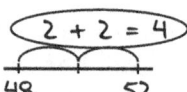
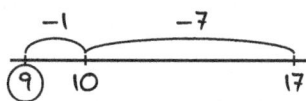

Numbers are close together –
find the distance/difference.

Numbers are far apart –
remove.

TIPS FOR MODELING STRATEGIES ON OPEN NUMBER LINES

When representing student thinking using open number lines, remember the following:

- Don't call a number line a strategy. A number line is a model.

- If you are drawing multiple number lines, align the number lines so that numbers are lined up vertically, so the relationships match.

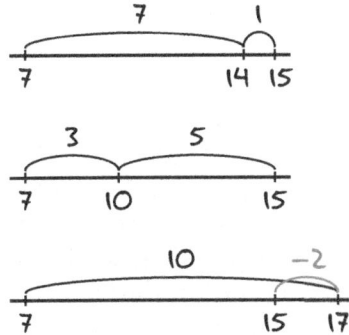

- Draw the jumps as proportional as you can. Do not overemphasize this with students. Just encourage them to draw big jumps longer than short jumps.

- Ask students where they land to motivate reasons for jumping, "Where did you land when you made that jump of 2? Ah, to 10, that's a nice friendly number" or "Where did you land when you jumped 10?"

- Sometimes beginning modelers will just draw jumps and not the line. *Draw the line.*

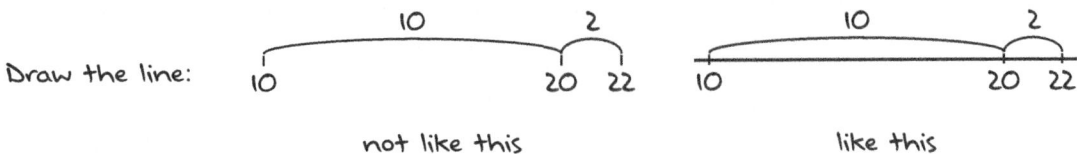

- On number lines, you can use + and − symbols to help students feel action and also to help students be able to represent their own thinking. Don't demand that students use + and − symbols unless you cannot figure out what a student is doing. Sometimes do not use the + and − symbols, so that students see distance and not just action. You can also use a different color to show the jump back to help students see what is happening.

SPLITTING

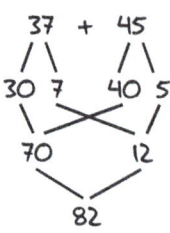

The splitting model is used to show how a student chooses to decompose and compose numbers. This is a fine beginning model for addition, but eventually you want to progress to using open number lines and equations. You could choose to never use the splitting model for addition. Either way, do not use the splitting model for subtraction—it's just not worth it. If a student is using it, feel free to help them use it correctly, but again, try to nudge everyone to the more versatile open number lines and equations.

EQUATIONS

Equations use the numeral symbols, operations, and equal sign to represent relationships. Equations can be used to represent thinking, but they are not a spatial model. For this reason, use them in conjunction with physical and spatial models for a long time before expecting students to use equations well.

$$62 - 28$$
$$= (62 - 20) - 8$$
$$= 42 - 8 = 34$$

> ### FREQUENTLY ASKED QUESTIONS
>
> **Q:** If equations are the model that students will eventually use most, why don't we just skip all the other ones and start with equations?
>
> **A:** Most students will not be able to make sense of all the underlying connections and relationships without the spatial models like a number rack and open number line. Remember, the purpose of math class is developing mathematical reasoning, not just getting answers.

OUR MODELING FRAMEWORK

You may have noticed from the examples of Problem String enactments in this book that we promote a certain framework for using models and modeling to help students develop mathematical reasoning:

- The teacher facilitates tasks imbued with important mathematical patterns.
- Students notice patterns and relationships while solving problems.

- The teacher elicits students' thinking and reasoning. The students share their strategies.
- The teacher represents that thinking with a model (could be a manipulative, visual model, equation).
- The students have the opportunity to realize that their mental actions can look like the models/representations, "When my brain does this, it can look like that." This helps students gain clarity on the patterns and relationships and how to use them.
- As students become familiar with the models and associated relationships, they begin to use those models as tools to reason with, in order to reason (Fosnot & Dolk, 2001).

Conclusion

The distinction between models and strategies is essential to teaching math effectively. Models provide the means of making student thinking visible. This visibility is a crucial step in creating the high-density doses of patterns students require to develop mathematical reasoning. With the correct use of models and modeling in math class, math is truly figure-out-able!

Discussion Questions

1. Which meaning of *model* in mathematics is the most familiar to you? Which is the least familiar?
2. How would you describe the difference between a *model* and a *strategy*? A *strategy* and an *algorithm*?
3. What is the difference between *tools to build relationships* and *tools for computation*?
4. Why are some models not good as tools for computation? What does it have to do with counting one by one?
5. How can you use an open number line as a tool to build reasoning and as a tool for computation?

CHAPTER 10

Moving Forward

It was the second day of our two-day in-person workshop where we filmed for our online Building Addition for Young Learners workshop (Math Is Figure-Out-Able, 2025). K–2 teachers and leaders had been diving into the same kinds of things you've been learning in this book. They were actively participating in Problem Strings and analyzing video of students working through problems.

At this point, teachers had just finished working together in randomly chosen groups at vertical nonpermanent surfaces (Liljedahl, 2021) on an adult-ish task, where we asked them to grapple with our base ten number system in a way that most had not before. They were determining how to best pack pencils from a warehouse, coordinating multiple levels of units (Hackenberg et al., 2016). We were having a Math Congress to clarify and build on their learning.

We had designed the task specifically to nudge teachers toward two distinct ways to solve one of the problems. In the comparison of the two methods, teachers were realizing that math class has a different goal than getting answers.

Katelyn said, with some emotion, "So it was just interesting how our numbers were the same because we figured it both ways. But it was so much easier the first way when we were subtracting, compared to the one that was a little bit easier to add. It was just super interesting. *And it really just made you familiar with the numbers and the relationships between them.*"

And that is the point. Making your brain familiar with the relationships, creating mental pathways and maps, and using them. More and more things ping, and we can do more and more real math.

MENTOR MATHEMATICIANS

Our role as teachers is to help the student mathematician develop their mathematical reasoning. Mentor the student as a math-er, helping them make the mental connections and pathways.

When we tell a student their answer is wrong, they've counted incorrectly, or they've messed up the steps, so here is the right answer, count this way, or do these steps like this, well, we might get correct answers, but the student's logical *working out* hasn't progressed. Right answers are necessary but not sufficient.

FREQUENTLY ASKED QUESTIONS

Q: But, Pam, are you saying we shouldn't correct students' wrong answers?

A: By using an asset perspective on what students are bringing to the work, and focusing on how you can help them develop forward, students will naturally correct their own wrong answers because as they are justifying their thinking, they will often catch their errors in the process. They will be more motivated to keep working and advancing their own reasoning rather than being defeated by a wrong answer.

You've just learned all about Counting Strategies and building Additive Reasoning K–2. It's time to implement. Where should you start?

> **TIP**
>
> It's better to start with a realistic goal that you can accomplish than an overambitious goal that sets you up for failure.

WHERE TO START

Think of a continuum of implementation: from building your own real math-ing to teaching a full-fledged *math is figure-out-able* classroom.

No matter where you are on this continuum, consider the following:

- Give yourself permission to not be perfect. Try things, learn from your experiences, try again with new understanding. Repeat.
- Focus on students' reasoning, not answers. Be intensely interested in how they are thinking. Ask questions to help them clarify their thinking and nudge them toward more sophisticated thinking, but don't try to replace their thinking. "Let me get clear on how you're reasoning.

Like this? Or that? Ah, we can represent that like this. Now that you can see your thinking, do any other ideas occur to you? Now that we're both clear on your reasoning, does Juan's strategy make sense?" Your goal is not answer-getting. Your goal is building mathematical reasoning.

- Work with a trusted colleague or coach as much as possible. Encourage each other. Share successes, failures, wonderings, and questions. Establish check-in times to hold each other accountable.
    - If you do not have this person or team, join the Math Is Figure-Out-Able Facebook Teacher Group [https://qrs.ly/9xgl27g] and our Journey teacher coaching system [https://qrs.ly/fhgl27f]. If you are a leader or coach, join our leader coaching program [https://qrs.ly/fhgl27f].
    - Build your numeracy. Work to solve problems using what you know. Learn the major counting principles and additive strategies and models. Take a Building Powerful Mathematics online workshop, focusing on your own mathematics: https://qrs.ly/ebgl27h

IF YOU'RE READY TO SHIFT

You're thinking that math is figure-out-able, and you want to begin shifting your teaching to make math more figure-out-able in your classroom. You have bandwidth to try some things, to get your feet wet, to learn and keep learning.

> **TIP**
> Read through each of the following possibilities. You will find helpful things to consider in each one.

Start by doing these things:

1. Take a Math is Figure-Out-Able miniworkshop. These two-hour online workshops are structured to give you manageable places to start. There's a good chance you can get your administration to pay for it. Find out more here: https://qrs.ly/plgr1ig

2. Carve out some time and space, at least *once a week*. You might decide to choose the end of the day on Tuesday, the beginning of math on Wednesday, or to replace your read-aloud time on Friday. Decide that during that once-a-week time, you will deliberately try lessons and routines to make math more figure-out-able.

3. During that time, facilitate a Problem String. Choose a Problem String from this book or from our grade level Problem String books (https://qrs.ly/5ngl27b), and give it a shot with your students. It may or may not have anything to do with what you are currently teaching–don't stress too much trying to decide on the perfect string; just get started.

Tell students you're all going to do something cool, math-ing the way mathematicians math.

4. When you have done a few Problem Strings, try a Count Around. Remember the important part is not the counting, it's the pattern finding and pattern discussing.

5. Play I Have, You Need at the beginning of math time or during random breaks as often as possible. Ask students about their strategies. Make it game-like.

6. Begin to notice where students could use reasoning to solve problems in your math times and elsewhere. Wonder what it would be like to handle those instances as figure-out-able.

7. Build your own numeracy by participating in #MathStratChat on your favorite social media platform.

> **TIP**
> Carve out *time* so that you make sure the transition happens. Tell a colleague or coach and work together to keep that time reserved.

IF YOU'RE READY TO TURN

If you're more in the middle of the continuum and you're ready to try, do, and plan more, consider these suggestions.

1. Take a Math Is Figure-Out-Able Building Powerful Mathematics workshop. These asynchronous online workshops take the Problem String video walkthroughs part of this book to an entirely different level. Meant to be experienced over a period of sixteen weeks, these workshops will set you up with the experience to give your students their best year in math yet. Find out more here: https://qrs.ly/chgl279

2. Carve out time and space *every day*. You might choose the beginning of math time or the end. You might choose to use that awkward fifteen-minute gap between science and lunch. Reserve this time for doing lessons and routines to make math more figure-out-able.

3. During that time, facilitate a Problem String. You could choose strings to (1) review prior constructed strategies, (2) get ideas percolating for your current unit of study, or (3) focus on constructing a strategy or a model.

4. Plan to use sequences of Problem Strings to gradually build toward more complexity and sophistication, helping students own the relationships more and more so that the strategies become natural outcomes. Work with a particular strategy for a couple of strings, put some words to the strategy, and do another string or two

> **TIP**
> Carve out *space* so students become clear that during that time their job is to reason mathematically. Ask students for a private response signal like a discrete thumbs up. Work on responding with a neutral response so that students learn to justify their reasoning. Use this space to practice your high leverage teacher moves that promote math-ing.

where you anchor the learning. Then switch your focus to a new strategy for a bit; then cycle back through a previous strategy.

5. About once a week, facilitate a different routine, like a Count Around or play a mathematically rich game (see Chapter 8).

6. Play I Have, You Need often and in short bursts. After playing for a while, discuss strategy. Play more. Compare strategy. Keep playing. After some time, switch so that sometimes you are doubling with students, eliciting their thinking. After students have doubled for a long time, ask them about halving numbers.

7. Take active note of where students could use reasoning to solve problems in your math times and elsewhere. Bring those instances to students' attention. "This problem reminds me of the ones we solved yesterday in the string. I wonder if that strategy could be helpful here." "We've seen numbers like this before, hmmm." or "What do you know that you could use to help here—this feels familiar?"

8. Use smudge problems (see Chapter 8) to encourage students to analyze, consider, and discuss relationships and patterns that lead to strategies.

In addition to those considerations listed previously for those Ready to Shift, consider the following:

- Committing to carving out time and space every day means that you will get to it on average four to five days a week. That's okay. Keep working at it. When the field trip happens, roll with it, and play I Have, You Need on the bus!

- Plan to use a purposeful sequences of Problem Strings, and do several in a row, but do not feel like you have to have everyone on the same page at the end of a string. Keep everyone learning and growing by letting them solve the problems how they want to, while you encourage them to use their best thinking. Keep the focus on the target strategy.

IF YOU'RE READY TO DIVE IN

If you have the bandwidth, and you're ready to dive in and make your math teaching figure-out-able, consider all the previous suggestions. Also, plan your math is figure-out-able instruction with these guidelines:

1. Sign up for the Math Is Figure-Out-Able Journey coaching program: coaching and QA sessions, access to the Problem String Hub with more than 60 videos of Problem Strings in real classrooms (and more added all the time), as well as miniworkshops. More information is available here: https://qrs.ly/qbgl27a

2. Divide your math teaching time into roughly a $\frac{1}{3} - \frac{2}{3}$ split.

3. During the $\frac{1}{3}$ block, review, preview, practice. This looks like Problem Strings, other instruction routines, intentional games, and assessment interviews. You might not have this time block during your regularly scheduled math time. You might create it during one of those funny gaps in your schedule, like the odd fifteen minutes between specials and lunch or the last twenty minutes of the day.

4. During the $\frac{2}{3}$ block, focus on the scope and sequence for your year. This looks like sequences of Problem Strings to get ideas percolating that lead into Rich Tasks/Math Congresses that are followed by Problem Strings to cinch a major concept, model, or strategy. Interspersed are other instruction routines and games to fill out the sequence.

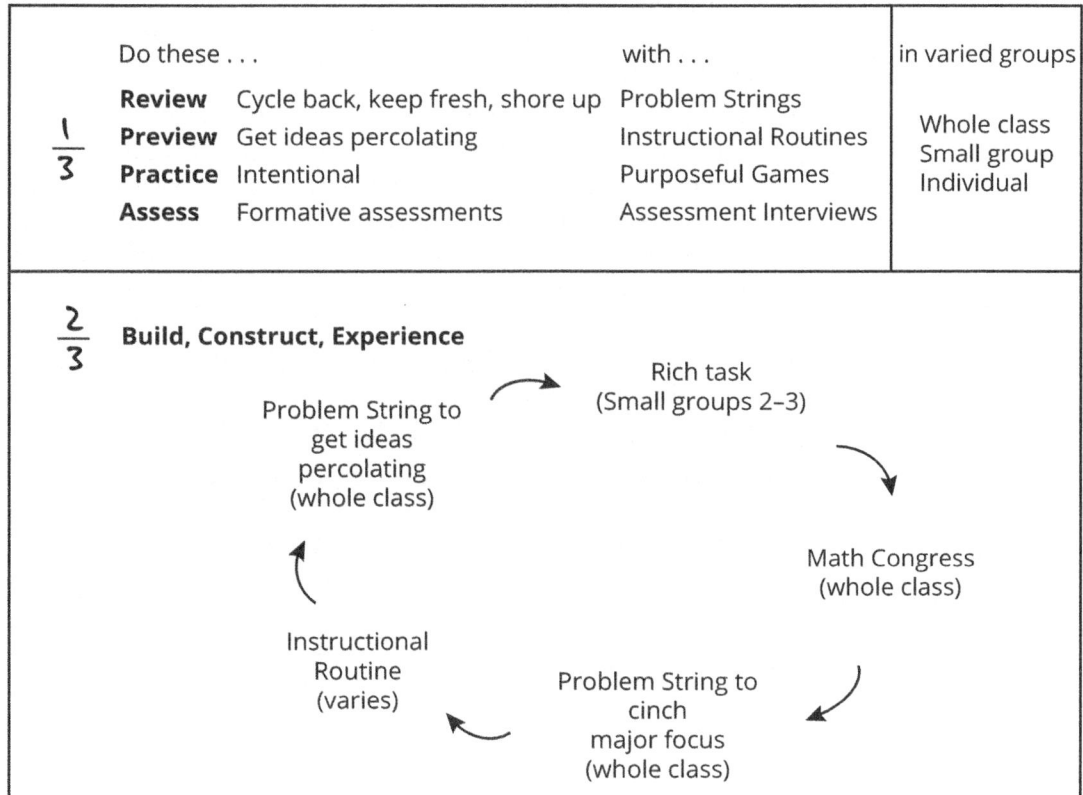

In addition to those considerations listed previously for those Ready to Shift and Ready to Turn, consider the following:

- A Rich Task can be shorter or longer or can also be a sequence of tasks, interspersed with Problem Strings and other routines.

- Other instructional routines might sometimes take the place of Problem Strings to get ideas bubbling up in class or to help cinch things.

- Delay sending work home. Send home things that students have already worked out with you and gained confidence. Then, when that stuff comes home, students will not be overly frustrated, and parents will not feel they need to step in to fix things with algorithms or rote-memorization. Games are a great choice for homework, if homework is required.

FREQUENTLY ASKED QUESTIONS

Q: Are you advocating that everyone wins, which really means that you are dumbing down math class so that everyone can keep up?

A: This is not about a trophy for everyone. We should advocate for excellence without leaving anyone behind; we should not mistake algorithm-mimicking for excellence. Helping students develop from where they are is the best way to support all students and challenge all students. By developing mathematical reasoning, we get more students doing more real math-ing, not less.

Q: Our school has tried so many new things. How do we make this work? How do we make it last?

A: Build support from the ground up. Share your successes with your colleagues—tell them the way you and your students are reasoning your way through problems. Start small, with those colleagues who have bandwidth. Help each other, and reflect your enthusiasm so that your success is contagious. Help your leaders understand what you are doing. Invite them into your classroom to hear your students developing and using mathematical reasoning before their eyes. If you are a leader, join our Journey Leader coaching program for more leader support.

Conclusion

Math is figure-out-able, and we can teach it that way!

Discussion Questions

1. Where are you on an implementation continuum? Ready to shift, to turn, or to dive in?
2. What is a first thing that you will try now that you are done with this book?
3. Where do you want you and your classroom to be one year from now? What's your goal?
4. Is there anything holding you back from starting tomorrow? How can you remove that barrier?
5. How will you get support? How can you leverage existing teacher support that is available to you? How can you seek support for your colleagues?

References

PREFACE

Crayton, D. (2026). *Readers read. Writers write. Mathers math! Bridging the gap between literacy and mathematics.* Corwin.

Harris, P. W. (2024). *Numeracy problem strings: Second grade.* Math Is Figure-Out-Able.

Harris, P. W. (2025a). *Developing mathematical reasoning: Avoiding the trap of algorithms.* Corwin.

Harris, P. W. (2025b). *Numeracy problem strings: First grade.* Math Is Figure-Out-Able.

Harris, P. W. (2025c). *Numeracy problem strings: Kindergarten.* Math Is Figure-Out-Able.

Liljedahl, P. (2021). *Building thinking classrooms in mathematics, Grades K–12.* Corwin.

CHAPTER 1

Crayton, D. (2026). *Readers read. Writers write. Mathers math! Bridging the gap between literacy and mathematics.* Corwin.

Harris, P. W. (2025). *Developing mathematical reasoning: Avoiding the trap of algorithms.* Corwin.

Jensen, E., & McConchie, L. (2020). *Brain-based learning: Teaching the way students really learn.* Corwin.

Lamon, S. (2020). *Teaching fractions and ratios for understanding: Essential content knowledge and instructional strategies for teachers.* Routledge.

Tanton, J. (2019). *Episode 6: Looking for joy in mathematics: An interview with James Tanton* [podcast]. Make Math Moments. https://makemathmoments.com/episode6/

CHAPTER 2

Chang, M., & Fosnot, C. (2025). *Beads and shoes, making twos: Extending number sense.* New Perspectives on Learning.

Clements, D. H. (1999). Subitizing: What is it? Why teach it? *Teaching Children Mathematics,* 5(7), 400–405. https://doi.org/10.5951/tcm.5.7.0400

Clements, D. H., & Sarama, J. (n.d.). *Learning and teaching with learning trajectories.* https://www.learningtrajectories.org/

Cobb, P. (1997). Instructional design and reform: A plea for developmental research in context. In M. Beishuizen, K. P. E. Gravemeijer, & E. C. D. M. van Lieshout (Eds.), *The role of contexts and models in the development of mathematical strategies and procedures* (pp. 273–89). Utrecht University. https://www.fisme.science.uu.nl/publicaties/literatuur/1997_Beishuizen_0-312.pdf

Fosnot C. (2025). *Bunk beds and apple boxes.* New Perspectives on Learning.

Fosnot, C. T., & Dolk, M. (2001). *Young mathematicians at work: Constructing number sense, addition, and subtraction.* Heinemann Educational Books.

Gelman, R., & Gallistel, C. (1978). *The child's understanding of number.* Harvard University Press.

Hackenberg, A. J., Norton, A., & Wright, R. J. (2016). *Developing fractions knowledge.* Sage.

Kamii, C. (1985). *Young children reinvent arithmetic.* Teachers College Press.

Piaget, J. (1965). *The child's conception of number.* Routledge.

CHAPTER 3

Carpenter, T. P., Fennema, E., Franke, M. L., Levi, L., & Empson, S. B. (2014). *Children's mathematics: Cognitively guided instruction* (2nd ed.). Heinemann.

Fosnot, C. T., & Dolk, M. (2001). *Young mathematicians at work: Constructing number sense, addition, and subtraction.* Heinemann.

MacCarty, K., Ellemor-Collins, & Wright (2025). On track to numeracy: A framework & tools for guiding classroom number learning. [Unpublished manuscript].

CHAPTER 4

Fosnot, C. T., & Dolk, M. (2001). *Young mathematicians at work: Constructing number sense, addition, and subtraction.* Heinemann Educational Books.

Institute of Education Sciences. (2024). *What does it mean to think additively.* https://ies.ed.gov/ncee/edlabs/infographics/pdf/REL_SE_What_Does_it_Mean_to_Think_Additively.pdf

CHAPTER 6

Harris, P. W. (2025). *Developing mathematical reasoning: Avoiding the trap of algorithms.* Corwin.

CHAPTER 8

Fernandez, C., & Yoshida, M. (2004). *Lesson study: A Japanese approach to improving mathematics teaching and learning.* Lawrence Erlbaum.

Fosnot, C. T. (2024). *Measuring for the art show: Addition on the open number line.* New Perspectives on Learning, https://newperspectivesonlearning.com/products/measuring-for-the-art-show-addition-and-subtraction-on-the-open-number-line

Fosnot, C. T., & Dolk, M. (2001). *Young mathematicians at work: Constructing number sense, addition, and subtraction.* Heinemann Educational Books.

Fosnot, C., & Dolk, M. (2001–2002). *Young mathematicians at work.* Heinemann.

Harris, P. W. (2024). *Numeracy problem strings: Second grade.* Math Is Figure-Out-Able.

Harris, P. W. (2025a). *Developing mathematical reasoning: Avoiding the trap of algorithms.* Corwin.

Harris, P. W. (2025b). *Foundations for strategies: Single digit addition and subtraction.* hand2mind

Liljedahl, P. (2021). *Building thinking classrooms in mathematics, Grades K–12.* Corwin.

Liu, N., Dolk, M., & Fosnot, C. (2024). *Organizing and collecting: The number system.* Perspectives on Learning.

Zull, J. E. (2011). *From brain to mind: Using neuroscience to guide change in education.* Stylus.

CHAPTER 9

Fosnot, C. T., & Dolk, M. (2001). *Young mathematicians at work: Constructing number sense, addition, and subtraction.* Heinemann Educational Books.

Gravemeijer, K. P. E. (1999). How emergent models may foster the constitution of formal mathematics. *Mathematical Thinking and Learning, 1*(2), 155–77.

Harris, P. W. (2023). *The most important numeracy strategies.* Math Is Figure-Out-Able.

Harris, P. W. (2024). *Major numeracy strategies* [ebook]. Math Is Figure-Out-Able.

Harris, P. W. (2025). *Foundations for strategies: Single digit addition and subtraction.* hand2mind.

CHAPTER 10

Hackenberg, A. J., Norton, A., & Wright, R. J. (2016). *Developing fractions knowledge.* Sage.

Liljedahl, P. (2021). *Building thinking classrooms in mathematics, Grades K–12.* Corwin.

Math is figure-out-able. (2025). *Building addition for young learners online workshop.* Retrieved May 17, 2025 from https://www.mathisfigureoutable.com/bayl

Index

Add 10 and Adjust strategy
    Doubles to Add strategy, 87–88
    implications, 96–97
    models, 93–94
    Problem String, 94–96, 94 (table)
    single-digits, 93
Addition (within 20), 69–77
    Counting Strategies, 78, 78 (figure)
    development, 81–82
    Doubles to Add strategy. *See* Doubles to Add strategy
    Get to 10 strategy, 83–87
    models, 84
    single-digits, 97–100
    strategies, 79–80, 79 (figure)
Additive reasoning, 13–15, 13–15 (figure). *See also* Addition (within 20)
Adult-ish task, 279
Algebraic reasoning, 15–16, 16 (figure)

Building Addition for Young Learners workshop, 279
Building Powerful Mathematics, 282
Building relationship models
    base 10 materials, 271
    difference/distance and removal, 274
    equations, 276
    10-frames, 269–270
    hundreds charts, 270–271
    number path, 270
    number racks, 272–273, 273 (figure)
    numerical and spatial, 268
    one-to-one counters, 269
    open number line, 273–274, 275
    splitting, 276

Cardinality, 25–26
Children's Mathematics: Cognitively Guided Instruction (2014), 50
Constant difference, 80
Constant Difference strategy
    equivalence, 192
    implications, 195–196

    models, 193, 193 (table)
    Problem String, 193–195, 194 (table)
    subtraction, 192
Contexts for Learning Mathematics, 206
Count Arounds, 142, 152, 244–245, 283, 284
Counting
    v. Counting Strategies, 22
    decades, 41–42
    development, 36
    foundations of number. *See* Foundations of number
    K–5 teachers, 21
    mathematical reasoning, 21, 21 (figure)
    number sequence, 37–41
    strategies. *See* Counting Strategies
    student interview, 42
Counting backward, 33–34
Counting principles, 22
Counting Strategies, 12, 12 (figure)
    Additive Reasoning, 13, 78
    bedrock, 12
    classroom activities, 60
    early, 46–48, 59
    mathematical reasoning, 45 (figure)
    objects/numbers, 60–61, 61 (table)
    on/back, 49–50, 61–62, 62 (table)
    problem types. *See* Problem types
    spatial reasoning, 15
    variations, 60

Decades, 41–42
Digital sums, 203
Dolk, M., 203
Double-digit addition
    accumulation strategy, 130
    developing multidigit addition, 131–133
    Human Calculator, 129
    Problem String, 230–233
    second-grade teacher, 130
    Splitting by Place Value. *See* Splitting by Place Value

Doubles to Add strategy, 88–89
   implications, 92
   models, 89–90
   Problem String, 90–91, 90 (table)

"Echo" strings, 238

Figure-out-able classroom, 281, 282, 283
Follow-up Task, 227–230
Fosnot, C., 233
Fosnot, C. T., 203
Foundations for Strategies: Single Digit Addition & Subtraction, 246
Foundations of number
   cardinality, 25–26
   conservation of number, 30–31
   counting backward, 33–34
   decision making, 23
   design environments, 23
   fine-grained nuances, 24
   hierarchical inclusion, 28–30
   one-to-one correspondence, 27
   one-to-one tagging, 28
   organization and keeping track, need for, 32
   subitizing, 24–25, 24 (figure)
   synchrony, 27
   unitizing, 34–35, 35 (figure)
Friendly Number, 80, 112, 131
Functional reasoning, 13–15, 13–15 (figure)

Games, 250

Hint cards, 251–252
Human Calculator, 129

Instructional routines
   analyzing strategic smudges, 246–247, 246 (figure)
   Count Arounds, 244–245
   on the fly, 247–250
   I Have, You Need, 241–242
   variations on I Have, You Need, 242–244, 243–244 (table)

K–2 teachers, 3, 14, 24, 279
Kindergarten/first-grade teacher, 12, 13

Making thinking visible, 236
Math Congress
   book baskets, 215–216
   Class Congress, 218

figure out place value, 221
Measuring for the Art Show, 223–227
Paper Bags, 214
pattern, 220
Problem Strings, 285
purpose, 213
sequence of tasks, 220–221
students' prior experience, 214
Mathematical reasoning
   additive reasoning, 13–15, 13–15 (figure)
   algebraic reasoning, 15–16, 16 (figure)
   continuum of implementation, 281–282
   counting strategies, 12, 12 (figure)
   digital sums, 203
   figure-out-able, 282
   functional reasoning, 13–15, 13–15 (figure)
   hierarchy, 11
   mentor, 280–281
   multiplicative reasoning, 13–15, 13–15 (figure)
   sequencing tasks. *See* Sequencing tasks
   sophistication, 11, 12, 12 (figure)
   spatial reasoning, 15–16, 16 (figure)
   statistical reasoning, 15–16, 16 (figure)
Measuring for the Art Show
   Follow-up Task, 227–230
   Math Congress, 223–227
   New Perspectives Online, 221
   Problem String, 230–233
   Rich Task, 222–223
   second-grade class, 221
Mentor mathematicians, 280
Middle- and high-school mathematics, 14
Mini-Congress, 211–213
Modeling
   Add a Friendly Number, 256
   effective uses, 255
   framework, 276–277
   Get to a Friendly Number, 256
   professional development sessions, 253
   safe learning environment, 256
   second-grade teacher, 253
   strategy, 254
Models
   building relationships. *See* Building relationship models
   calendar scenario, 265
   common usages, word, 260–261
   cubes, 266
   Jodi's students' experience, 265
   Kindergarten, second-grade classes, 267–268

mental maps, 264
principles, 267
representation of the situation, 263
representation of their thinking, 266
representing thinking, 261–263
v. strategies, 256–260
tools for thinking, 264, 266
use of manipulatives, 264
See also Modeling
Multidigit subtraction, 165–174
comparison, 196–199, 196 (table)
Constant Difference, 176
development, 175–177
distance/difference strategy, 190–191
Remove a Friendly Number Over, 176
Remove a Friendly Number Strategy. *See* Remove a Friendly Number Strategy
Removing by Place Value, 177–178
Split by Place Value, 176
Multiplicative reasoning, 13–15, 13–15 (figure)

New Perspectives Online, 221
Number sense, 3, 23, 150, 272
Numeracy Problem Strings, 233, 234

One-to-one correspondence, 27
One-to-one tagging, 28

Partial sums, 137, 138
Problem Strings, 62–65, 63 (table)
Add a Friendly Number Over Strategy, 150, 151 (table)
Add a Friendly Number Strategy, 141, 141 (table)
anatomy, 236–239
Constant Difference Strategy, 193, 194 (table)
Count On and Count Back, 62–65
Doubles to Subtract Strategy, 116–117 (table)
first-grade teacher, 5
games, 285
Get to a Friendly Number Strategy, 146, 146 (table)
Give and Take Strategy, 155, 156 (table)
great teacher first step, 239–240
instruction routines, 285
math forward, 236
miniworkshops, 285
number rack, 65, 90
purposeful sequence, 233, 234, 284
Ready to Shift, 286

Ready to Turn, 286
Remove a Friendly Number Over Strategy, 188, 188 (table)
Remove a Friendly Number Strategy, 180, 180 (table)
Remove to 10 strategy, 107
set of related problems, 234
spatial and operational relationships, 235–236
Splitting by Place Value Strategy, 136, 136 (table)
support students, 234–235, 235 (figure)
visual and numeric relationships, 235–236
Problem types
action/no action, 55–56
combining, 51, 52 (table)
comparison, 51, 54 (table)
number size, 58
part-part-whole questions, 51, 53 (table)
separating, 51, 53 (table)
solutions, 62
strings, 62–65, 63 (table)
students' ability, 50
unknown, 56–58

Remove a Friendly Number Over strategy, 187
implications, 190
models, 188
Problem String, 188–189, 188 (table)
Remove a Friendly Number Strategy, 179–180
implications, 182–183, 186
models, 180, 180 (table), 184
Problem String, 180–181 (table), 180–182, 184–186, 185 (table)
subtrahend, 183
Remove to 10 strategy
implications, 112
models, 109
problem string, 109–112, 110 (table)
single-digit number, 108
Removing by Place Value, 177–178
Rich Task, 222–223, 285

Sequencing tasks
bit-by-bit progression, 204
games, 250
helping students, 208
hint cards, 251–252
instructional routines. *See* Instructional routines

Japanese educators, 205
K–1 combination class, 206
landscape of learning, 204
Math Congress, 213–221
Measuring for the Art Show. *See* Measuring for the Art Show
mini Class Congress, 211–213
open access, 205
paper bag puppets, 206
Problem Strings. *See* Problem Strings
students working, 208–211
study–exemplar tasks, 205
Single-digit numbers, 81, 84, 97, 108, 120, 142, 152
Southeastern Regional Education Laboratories, 24
Spatial reasoning, 15–16, 16 (figure)
Splitting by Place Value
    Add a Friendly Number, 139–144
    Add a Friendly Number Over strategy, 149–153
    digit-focused algorithms, 134
    Get to a Friendly Number strategy, 144–149
    Give and Take strategy, 153–158
    implications, 137–138
    major addition strategies, 158–162
    models, 134, 134 (table)
    multidigit addition, 134
    Problem String, 135–137, 136 (table)
Statistical reasoning, 15–16, 16 (figure)
Subitizing, 24–25, 24 (figure)
Subtraction (within 20), 103–105, 104 (figure), 105 (figure)
    developing, 105–107
    distance/difference strategy, 124–125
    doubles to subtract strategy, 114–119, 116–117 (table)
    Remove 10 and Adjust Strategy, 119–124
    Remove to 10 strategy. *See* Remove to 10 strategy
    single-digit subtraction strategies, 125–126
Synchrony, 27

Tanton, J., 4
Teaching Children Mathematics (1994–2019), 203
Teaching mathematics, 3–10
    mathematical reasoning. *See* Mathematical reasoning
    purpose, 10
    strategies, 16–17

CORWIN

To help every educator help every student

We believe that every single student deserves a great education

We believe that knowing our impact is both a privilege and a responsibility

We believe that a fair, stable, and thriving society is built on education

Supporting Teachers, Empowering Students

Peter Liljedahl

Fourteen optimal practices for thinking that create an ideal setting for deep mathematics learning to occur.
Grades K–12

Peter Liljedahl, Maegan Giroux, Kyle Webb

Delve deeper into the implementation of the fourteen practices from Building Thinking Classrooms in Mathematics by focusing on the practice through the lens of tasks.
Grades K–5, 6–12

Rachel Lambert

Discover UDL for math, a way to design math classrooms that equips all students for meaningful and joyful math learning.
Grades K–8

Jo Boaler, Cathy Williams

Introduce data science to your students across disciplines with real-world stories and teacher testimonials to transform your classroom experience.
Grades K–8

Kimberly Rimbey, Katie Basham, Chryste Berda

Focus on making mathematics meaningful through multiple strategies and representations to help foster a love for mathematics in your students.
Grades K–6

Sherri Martinie, Jessica Lane, Janet Stramel, Jolene Goodheart Peterson, Julie Thiele

Build critical intrapersonal and interpersonal skills *through* mathematics to help all students grow the life-skills they'll carry forever.
Grades K–12

To order your copies, visit corwin.com/math

CORWIN Mathematics

Our research-based and high-quality content is written by trusted experts and provides clear pathways to helping all students gain access to rigorous mathematics learning; to learn to truly think, reason, collaborate, and fluently discuss mathematics; to form positive and strengths-based mathematical identities; and to see and use mathematics as a tool to effect change in their lives and communities.

Sue Chapman, Holly Burwell, Mary Mitchell

Focus on developing eight essential habits that support mathematical competence and confidence in students through a personalized, practice-based professional learning experience.
Grades K–5

Pamela Weber Harris

Guide students through domains of mathematical reasoning, from counting and adding strategies to more complex proportional and functional reasoning—without resorting to algorithms.
Grades K–12

John J. SanGiovanni, Dennis McDonald

Enhance your students' understanding and engagement in geometry, measurement, and data while also fostering a deeper connection between math and the real world.
Grades K–5

Vanessa "The Math Guru" Vakharia

Equip students to develop the skills they need to truly believe anything is possible, even a better relationship with math!
Grades K–12

Liesl McConchie

Reexamine what it means to have a positive math identity—and learn to use brain-based tools in a humorous and friendly way to build on a positive math identity for your students.
Grades K–12

Irina Lyublinskaya, Xiaoxue Du

Integrate AI literacy into K–12 classrooms, blending theory, practical lesson plans, and ethical considerations to empower students as critical thinkers.
Grades K–12

To order your copies, visit corwin.com/math

CORWIN Mathematics

In compliance with GPSR, should you have any concern about the safety of this product, please advise: International Associates Auditing and Certification Limited, The Black Church, St Mary's Place, Dublin 7, D07 P4AX Ireland EUAR@ie.ia-net.com

Printed and bound by CPI Group (UK) Ltd, Croydon, CR0 4YY
10/04/2026
02087355-0002